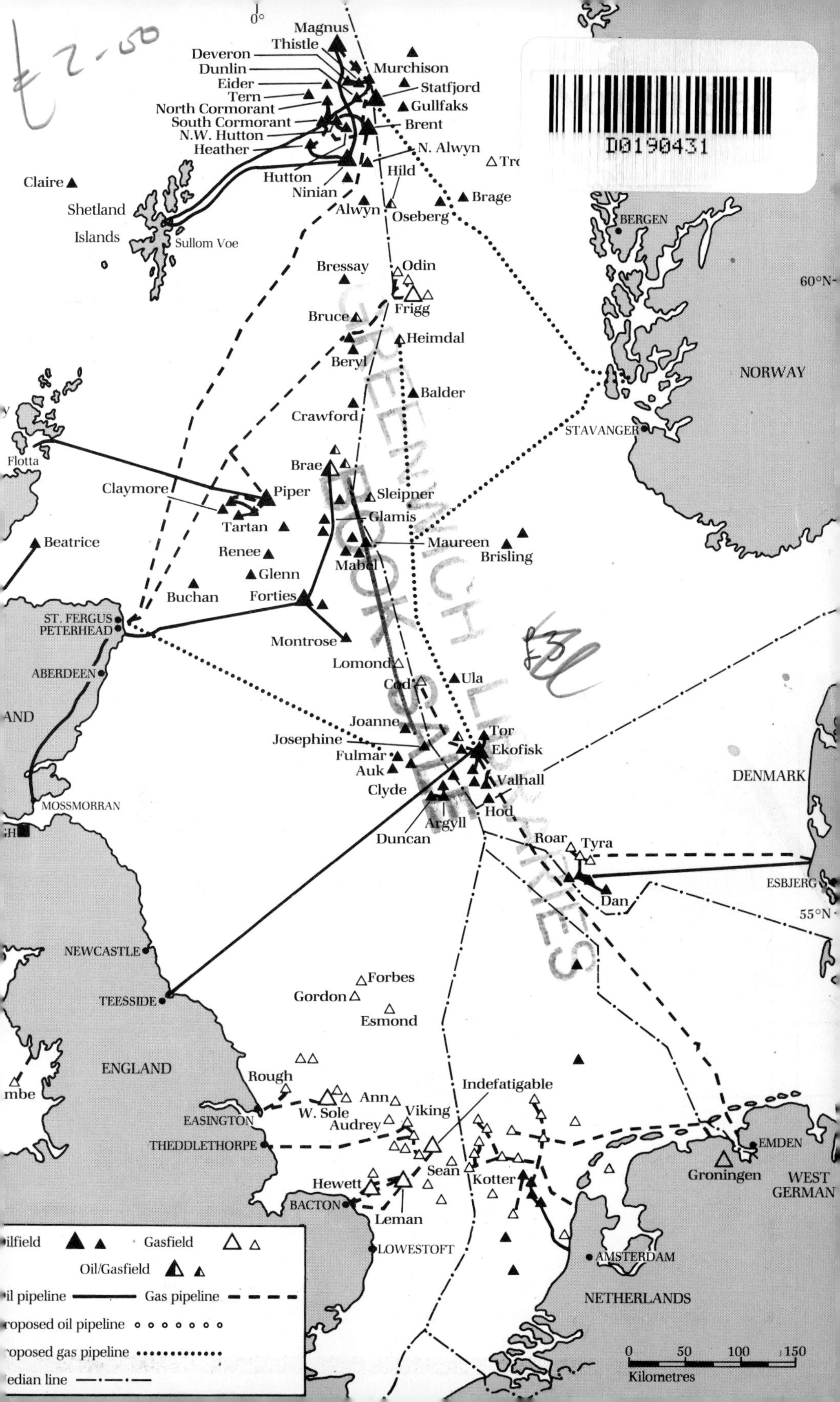

Magnus
Thistle
Deveron
Dunlin
Eider
Tern
North Cormorant
South Cormorant
N.W. Hutton
Heather
Murchison
Statfjord
Gullfaks
Brent
N. Alwyn
Hutton
Ninian
Hild
Alwyn
Oseberg
Brage
Claire
Shetland
Islands
Sullom Voe
BERGEN
Bressay
Odin
Frigg
60°N
Bruce
Heimdal
Beryl
NORWAY
Balder
Crawford
STAVANGER
Flotta
Brae
Claymore
Piper
Sleipner
Tartan
Glamis
Maureen
Brisling
Beatrice
Renee
Mabel
Glenn
Buchan
Forties
ST. FERGUS
PETERHEAD
Montrose
ABERDEEN
Lomond
Cod
Ula
Joanne
Josephine
Tor
Ekofisk
Fulmar
Auk
DENMARK
Clyde
Valhall
MOSSMORRAN
Argyll
Hod
Duncan
Roar
Tyra
ESBJERG
Dan
55°N
NEWCASTLE
Forbes
Gordon
TEESSIDE
Esmond
ENGLAND
Rough
Indefatigable
Ann
W. Sole
Viking
EASINGTON
Audrey
THEDDLETHORPE
EMDEN
Sean
Groningen
Hewett
Kotter
WEST
GERMAN
BACTON
Leman
Gasfield
Oil/Gasfield
LOWESTOFT
AMSTERDAM
Gas pipeline
NETHERLANDS
0
50
100
150
Kilometres

Offshore

Also by A. Alvarez

General
Under Pressure
The Savage God: A Study of Suicide
Life After Marriage: Scenes from Divorce
The Biggest Game in Town

Novels
Hers
Hunt

Criticism
The Shaping Spirit
The School of Donne
Beyond All This Fiddle
Samuel Beckett

Poetry
Lost
Apparition
Penguin Modern Poets No. 18
Autumn to Autumn and Selected Poems 1953–76

Anthology
The New Poetry (Editor, and Introduction by)

Offshore

A North Sea Journey

A. Alvarez

HODDER AND STOUGHTON
LONDON SYDNEY AUCKLAND TORONTO

The contents of this book originally appeared, in slightly different form, in *The New Yorker*.

Maps by Martin Lubikowski
Text illustrations by Bill Le Ferer

British Library Cataloguing in Publication Data

Alvarez, A.
Offshore: a North Sea journey
1. Offshore oil industry – North Sea – Employees 2. Offshore gas industry – North Sea – Employees
I. Title
627'.98 HD8039.0342N6

ISBN 0-340-37347-4

 Printed in Great Britain for Hodder and Stoughton Limited, Mill Road, Dunton Green, Sevenoaks, Kent by St Edmundsbury Press, Bury St Edmunds, Suffolk. Photoset by Rowland Phototypesetting Limited, Bury St Edmunds, Suffolk. Hodder and Stoughton Editorial Office: 47 Bedford Square, London WC1B 3DP.

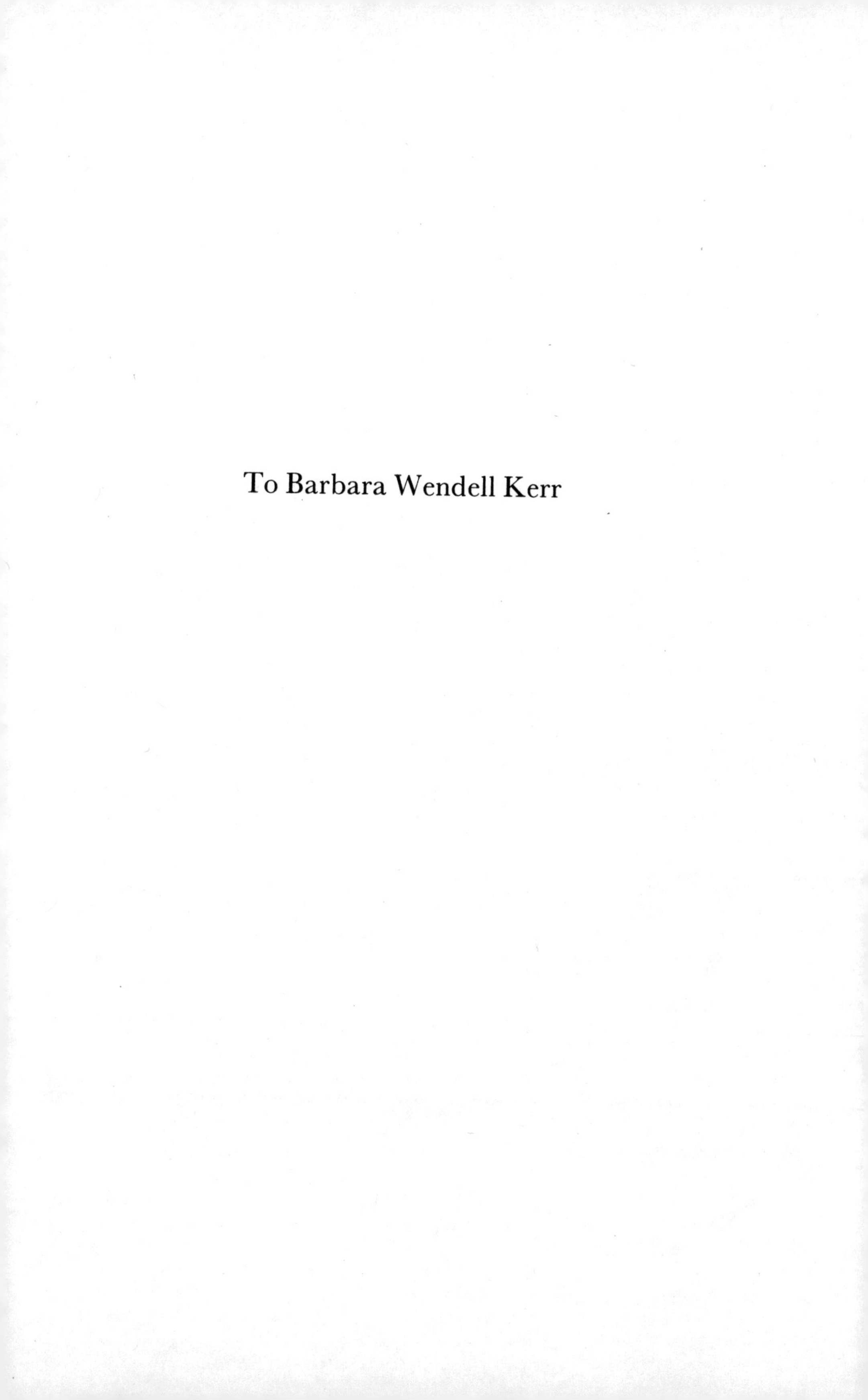

To Barbara Wendell Kerr

The notion of looking on at life has always been hateful to me. What am I if I am not a participant? In order to be, I must participate. I am fed by the quality in those who participate with me. That quality is something the men of the Group never think of – not out of humility, but because they do not stop to measure it . . . Each of these men is a web woven of his job, his trade, his duty . . . The presence of these men is dense, full of meaning, and it warms my heart. I am able to sit with them in silence . . . I do not mean to belittle the workings of the mind or the products of the intelligence. I admire a limpid intelligence as much as any man. But what is a man if he lacks substance? If he is a mere intellect and not a being?

Antoine de Saint-Exupéry, *Flight to Arras*

Acknowledgments

For the outsider, a journey to the North Sea oil fields is impossible without help from one of the operating companies. I went to Shell because Sean Williams, an old friend and climbing companion who works for the company, put me in touch with their Public Relations Department in London. Once they had checked me out, they were endlessly helpful in getting me offshore, answering my questions, and arranging for me to go wherever I wanted and meet whomever I asked to see. I owe a special debt of gratitude to Jim Arnott in London and to Alan Jacobs and Jean Roberts in Aberdeen. Between my first and second trips to the Brent field, George Band, another climbing friend, returned from abroad to Shell-Mex House as a director of Shell Exploration and Production (Shell Expro), smoothing my way even more and introducing me to several key figures in the development of North Sea oil. Although Band has now become director general of the prestigious United Kingdom Offshore Operators Association, he generously took time off from this demanding job to read the first draft of this book in meticulous detail and help eliminate a large number of mistakes. I am very grateful to him. I would also like to thank the people I talked to offshore, all of whom remained helpful and polite despite the pressures of their work and the often inappropriate moments I chose to badger them with questions.

I am grateful, above all, to William Shawn, the editor of *The New Yorker*, for his unequalled patience and generosity, and also for his willingness to trust me with an unlikely topic.

Introduction

Before I went offshore to the Brent field in 1983, I had had only three experiences of the North Sea, all of them unpleasant. In April 1948, with eighteen months to kill between school and university, I presented myself at Captain Watts's plush ships' chandlers shop in Albemarle Street and asked for a job on a boat. I do not now know why I imagined that a ships' chandlers – even one situated just off Piccadilly – would provide me with a ticket to life afloat. I do not even know why I imagined that I wanted to go to sea. Perhaps I had read too many schoolboy adventure stories. Or perhaps it was simply because I was eighteen years old and had never even crossed the Channel; like most young people in Britain after the war, all I really wanted was to get away.

Whatever the reason, I managed to con my way through the downstairs shop crammed with tackle and ropes and oilskins, up to the captain's private office. 'Go anywhere, do anything,' I announced, like an advertisement in *The Times* agony column.

Captain Watts had a fiery complexion and a fierce stare. 'Any experience?' he barked.

'Five seasons sailing on the Norfolk Broads.' It had seemed like a lot at school, where I had made much of my skill as a yachtsman. With the captain's framed certificates and photographs around me, I wasn't so sure.

Captain Watts rolled his eyes to the ceiling, then shuffled impatiently through some papers on the desk, waiting for me either to say something or go.

'Anywhere,' I repeated doggedly.

He selected a sheet of paper from the sheaf in front of him and studied it. 'If you can get yourself up to Scotland,' he said grudgingly, 'there's a ship in Port Seton we're bringing south next week. I suppose I could use another deck hand.'

'Thank you. Thank you very much.' I stared at my shoes, trying to work up my courage. I had never had a job before. 'About the . . . wages . . .'

'No wages. Experience is what you'll get. Valuable experience.'

'Thank you,' I repeated. 'Thank you very much.'

I borrowed some money from my bewildered parents and bought a one-way ticket to Edinburgh.

Port Seton is a few miles east of Edinburgh, on the blowy Firth of Forth. It was low tide when I arrived and only the ship's tall, old-fashioned funnel was visible above the dock. She was an eighty-foot steam drifter – I can no longer remember her name – built in Dundee in 1880, battered, rusty, ungainly – a 'dirty British coaster with a salt-caked smoke stack' straight out of Masefield. I climbed down the steel ladder from the dock and called, 'Hullo.' No response. A gull, perched on the seaward rail, peered at me with distaste. I went aft, down some steps, into a corridor, and called again. A muffled voice answered from the end of the passage. I opened the door to a cramped and stuffy saloon where a wolfish, middle-aged man and a heavily made-up woman were sitting at a table, drinking pink gin. They stared at me in amazement. I was wearing a tweed jacket, an Old Oundelian tie, and spotless cream-coloured corduroys.

'Captain Watts sent me,' I said.

Silence. They went on staring at me.

'I'm here to work.'

They stirred, blinked, as though coming awake.

'Then you'd better go and see the stoker. The hatch amidships. Leave your gear in the fo'c'sle.'

The fo'c'sle was small and curved, with six bunks and a stained table. The air was stale, pungent, and slightly rancid. I dumped my bag, jacket, and tie on a top bunk and went back on deck. A hatch was open on the seaward side of the boat. I climbed down into a narrow black hole, full of coal.

The stoker was a tiny, muscular Scot, covered with sweat and black with dust. He squinted at me through the gloom and said, 'Fuck me!' But by then I was getting used to the effect I had on strangers.

For the next three days I worked with him in the hole, shovelling coal as it was dumped, sack after sack, down the hatch. On the second day the wolfish man lit the boilers while the stoker and I laboured, and from then on the hole was like Bombay in a black dust storm. On the third day a line stuck in a block on the forward mast, high up beyond where the ladders gave out. I managed to monkey my way up and free it (any job in the fresh air was preferable to the black hole), and after that the wolfish man looked less depressed when he saw me.

On the fourth day I was given the afternoon off. I washed off most of the coal dust, put on my tweed jacket and Old School tie to offset the now black cords, took a bus into Edinburgh, and went straight to a barber. When he rinsed my hair the water ran like ink down the drain. By the time I got back to Port Seton, Captain Watts had arrived with the rest of the crew – a youth a little older and more experienced than myself, who ignored me, and four purposeful men who were working towards their Master Mariner's Certificate. We sailed the next morning on the dawn tide.

As I remember it, the boat began rolling the moment we slipped our moorings, and did not stop until we tied up again in London.

The four-hundred-mile journey took a week. We crawled down the east coast, always tantalisingly in sight of land. The sun shone but a cold wind blew relentlessly from the east, whipping the waves into white horses and slamming them into the shuddering old ship. 'She's a regular cow,' the stoker had warned me before we left. 'She'll roll in dry dock.' He had also said, of the wolfish man's wife, 'She's so tightfisted she wouldna' gee you a thick ha'penny for a thin un.' On both counts he was right. The old tub rolled all the way down the coast and went on rolling in the milder, more sheltered waters of the Thames estuary. Only when we steamed dead slow into Wapping 19 – fifty square yards of oily green water on which not even the debris stirred – did I seem to feel the uneasy motion slowly fading in her.

It had taken a day for seasickness to get me but by that time everyone except the stoker was ill. Even Captain Watts and the master mariners were vomiting over the side before we were six hours out of Port Seton. I watched them, amazed that they should have chosen a profession that made them suffer so much, although by the time my turn came they had recovered.

It was a miserable week but the night watches were the worst. The ship was so ancient that it had no electricity and was lit, instead, by acetylene lamps. (Hence that pungent, faintly rancid smell that permeated the fo'c'sle and every other enclosed space.) The most obnoxious of these lamps was in the binnacle housing the compass on the bridge. It was an overhead binnacle, glassed in on three sides, with the compass card at eye level as you held the brass-bound wooden wheel. But at night the little acetylene jet that illuminated it created confusing reflections in the glass, and in order to read the bearings you had to stick your nose right into the unglassed side, a few inches below the flaring jet. Twice I was up on the bridge on my own from midnight to four a.m., inhaling the stink of acetylene and trying to keep the ship on course while the waves thumped monotonously into her port beam and her bows heaved up and down. The first time I left her a mile off course, the second nearer three miles. After that, the master mariners got the midnight watch at the helm and I relieved the stoker.

When we finally passed Southend pier, sticking out like a pencil into the Thames estuary, and London became a smudge of smoke in the distance, I felt like a pilgrim sighting Rome. The rackety Underground branch line from Wapping to Whitechapel was civilisation itself – interior-sprung seats, a floor that vibrated but did not roll, and people, ordinary people, going to work. London seemed a natural place to be, man-made and, despite its sprawl, scaled to man. In the North Sea we had seemed intruders, existing precariously and under sufferance. What Delmore Schwartz – another city boy – called 'the fatal, merciless, passionate ocean' is not tolerant of people or their mistakes. The evil white horses walloping unforgivingly into our suffering old ship were a constant reminder of who was boss.

Captain Watts had been right when he said I would earn myself experience. I had learned that the sea was not my element, and one week's misery was a cheap enough price to pay.

It was twenty years before I went back to the North Sea, as crew on the yachts of friends, once crossing to Ostend, once to Rotterdam, both times without pleasure. The crossing to Ostend was an easy dawdle in mild weather, but by the time we had gorged ourselves on *moules marinière* and white wine the wind had dropped entirely and we had to chug back on the tiny auxiliary engine, with the sails hanging inertly. Although the sea was as flat and featureless as milk, the fumes from the engine made all of us sick. The journey to Rotterdam was also on the auxiliary engine, but for the opposite reason: a force-eight gale was blowing straight into our teeth, making it impossible even to hoist the sails. At some point on the second day, my elder son – then aged eleven – asked me how much longer it would be before we arrived. He was in the aft cabin, where he had been from the start, lying on a bunk and vomiting sporadically into a bucket.

'Some time yet, I'm afraid.'

'Two hours?'

I shook my head.

'Six hours?'

'More like two days,' I said.

Two large tears started from his eyes. He turned his face to the bulkhead and stayed like that until we docked. It was his first North Sea journey, and also his last.

To each his own element. The mountains, which I love, can be as hostile and unpredictable as the sea. They, too, do not accommodate themselves to the human scale, cannot be fooled, do not relent. But like all climbers, I enjoy heights and get a kick out of the sense of exposure. I also understand, more or less, what is involved and feel I can cope, more or less, if things go wrong. The sea, quite simply, scares me; its great depths of black water induce in me a kind of vertigo. I also find it monotonous, lacking the physical variety of the mountains, the constant ebb and flow of feeling on a climb – the build-up to the difficult pitches, the relaxation on the belay. Even under sail, in good weather or bad, ninety-eight per cent of the time at sea seemed to me rather boring. The other two per cent were as terrifying as being caught in an avalanche – a sense of total impotence, of being reduced to nothing at all by uncontrollable natural forces.

I suppose it was precisely because the North Sea gave me the jitters that I was fascinated by the idea of oil fields up in its northern sector, where colonies of people worked all year round in the unspeakable weather. It seemed not just another technical accomplishment but a kind of mystery, and the photographs of the installations rising improbably out of the water – their size and complexity and apparent clumsiness – only increased the strangeness.

The oil installations are strange in the same way as the awkward, seemingly patched-together contraptions NASA puts into orbit are strange. Compared with most products of high technology – aeroplanes, computers, even cars – they look so inept that it seems impossible for them to do the intricate jobs they were designed for. And the jobs, in turn, are so complex that, to the outsider, the ingenuity required to perform them seems like magic. Perhaps it is a question of language – or rather, the lack of it. The astronauts go into space and come back talking in the same cramped, polysyllabic clichés they might use if they had been to Topeka, Kansas. Not one of them has yet described what it looks and feels like out there.

As a result, the pictures and diagrams seem no more real than the illustrations in a science-fiction fantasy. Similarly, the oil installations exist slightly beyond the level of our imagination or our attention, performing their mysterious functions off somewhere in the north, as remote from ordinary reality as satellites.

As the oil companies constantly remind us, North Sea oil has a great deal in common with the space programme. Both are explorations of the unknown, and both are gigantic technological achievements, involving thousands of experts and vast expense, and extending the limits of scientific knowledge. But the size and complexity of the British operation touches the imagination less than its sheer improbability: a whole industrial province in the middle of nowhere, factories and airports and hotels rising out of the stormy waters on the sixty-first parallel, more than a hundred miles northeast of the Shetlands and nearly two hundred miles northwest of the nearest railhead – which happens to be Bergen, in Norway. The North Sea oil fields provide Britain with something it lost when it gave up its colonies – a frontier.

They have also taken over where the colonies left off, as a God-given source of wealth. Oil may not have made Britain rich again, but at least it has kept it solvent at a time when it seemed that nothing could prevent it from sliding towards the bottom of the European economic ladder, somewhere between Spain and Greece. Mercifully, the early estimates of the offshore oil reserves were sufficiently pessimistic to discourage the complacency that was once inspired by an empire on which the sun never sets. When the oil first began to come ashore it was greeted, by both politicians and press, as though it were little more than a brief remission from a terminal illness, a remission that might last for a decade or so, then leave us sicker than before. But as the economist Peter Odell has demonstrated in *The Future of Oil*, the calculation of oil reserves is notoriously unreliable, since it depends on constantly shifting economic, political, and technological factors, all of which encourage the producing companies generally to underestimate their reserves. So the decade has passed and the oil is still flowing. More than that, the industry now predicts that it will continue to flow, more or less undiminished, at least until the end of this century. Yet these predictions – published in September 1984 by the United Kingdom Offshore Operators Association, the official voice for all the companies involved in producing oil from British waters – were based almost entirely on what UKOOA calls 'relatively mature areas of the North Sea'. They take no account of the potentially large reserves that may lie in Britain's western waters that extend out into the Atlantic towards Rockall. In 1983, when I was researching this book, I talked to George Williams, the brilliant geologist whose professional flair and personal enthusiasm led to the oil strikes off Nigeria and Brunei, and, above all, in the East Shetland Basin. He said, 'My nightmare is to wake up in the middle of the next century, when oil is no longer the main source of man's energy requirements, and see from above huge undiscovered pools of oil in the western waters of the UK, and to hear the Almighty say, "I put it there and you were stupid enough not to look for it." '

The difficulties of producing oil from the Atlantic are prodigious, since the water is about ten times deeper than that of the East Shetland Basin – around five thousand feet rather than five hundred. But fifteen years ago, when no one had experience of drilling in waters deeper than two hundred feet, the problems of the North Sea appeared equally insuperable. 'Producing oil from five thousand feet of water is just a technical problem,' said George Williams. 'And technical problems are always solved. It's like war: if you want something urgently enough and are willing to pour in enough money, people are going to do it.'

That style of confidence, the scientist's assurance that any obstacle can be overcome with sufficient time and money, struck me as almost as strange as the North Sea project itself. Writers are not, by profession, in the solution business; their subject matter is doubt, the cracks in the fabric and points of weakness where character shows through, and whatever confidence they have comes from their precarious control of their medium. So going offshore brought me face to face with a different way of thinking as well as with a different world. Out in the Brent field and back in the Shell offices in Aberdeen I met people whose business it was to cope, to solve problems – engineering problems, geological problems, administrative and logistical problems – people who knew about drilling mud and computers and navigation and critical path analysis and cash flow, about aircraft and their maintenance, about transporting supplies and personnel, about the techniques of living and working on the bottom of the sea. Whether or not they were equally competent in their private lives was beside the point. For them, doubt was just another problem to be solved and they judged themselves as they were judged – pragmatically, by their performance. 'You are dealing,' said one of them, 'with real things and taking large risks.' To a desk-bound writer, that was the most irresistible of siren calls.

I have always been fascinated by hostile environments, and the northern North Sea is one of the most hostile on earth. But unlike the mountains, survival up there is not an end in itself. Nobody goes offshore for the restricted pleasure of being in an unforgiving place where simply getting by is its own challenge and reward. They are there to perform specific jobs, many of them extremely intricate and technical. And so involved are they in their work that the environment, hostile or not, scarcely impinges on them, except as a periodic frustration when storms or fog prevent them from leaving an installation or going out to one. Since a large number of offshore staff have served in the Royal or Merchant navies before going to work for the oil companies, that particular form of hostility is, anyway, something they have lived with so long that they barely notice it.

Because I was hoping to see the conditions at their worst, I arranged for my first trip offshore to be in early March, when the 'weather window' would not officially have been open a few years earlier – although now work goes on all year round. Even so, there were no storms while I was out there. There was fog, wind, rain, but nothing dramatic. When I went back in July there was a heat wave and the North Sea was as mild as the Aegean. I made a private arrangement with Ric

Charlton, then Shell's director of operations in Aberdeen, for him to call me if the barometer began to drop spectacularly, so that I could get myself up to Aberdeen and on to a helicopter to Brent. But Charlton was posted soon after to Malaysia and the call never came. I still feel disappointed to have missed the show.

While I was researching the book, people were continually asking me, in so many words, what a literary guy like me was doing in a joint like the Brent field. They seemed to be implying that the literary-intellectual world is somehow superior to all that practical stuff. Although that is not a belief I have ever shared, I felt I had to give my reasons, make my excuses, and foremost among them was the pleasure of being there and seeing it all for myself. This seemed the least convincing to literary intellectuals, but it happened to be the truth: my real motive for going offshore was simple, overwhelming curiosity. The diary I use prints at the foot of each page an aphorism for the day. Most are dreadful in a cosy way, but one stuck in my mind: 'Forty is the old age of youth, fifty the youth of old age.' I was fifty-two years old, my second youth was slipping away from me, yet I knew nothing at all about this product that I used every day, about this bonanza that had miraculously transformed the economy of my country, about this strange new frontier that had unexpectedly been added to overcivilised Europe. It started as a mystery, and although I finished knowing something about how oil is extracted from below the seabed and about how the extraordinarily complex installations work and about what it is like to live on them, none of that finally made the mystery any less. Instead, it changed the focus: the more I understood, the more mysterious and impressive became the ingenuity, perseverance, organisation, and sheer hard work that have made the North Sea oil province possible.

1

The electronic radio-clock by my bed in the Altens Skean Dhu Hotel near Aberdeen's Dyce Airport had a digital read-out and as many buttons as a Japanese hi-fi, but the alarm failed to go off at five-thirty. I woke in a panic at six and when I reached the lobby ten minutes later David Betts, the tall, pale young man who was to go with me on the three-hundred-mile trip to the Brent oil field, was pacing nervously up and down in front of the reception desk. When we climbed into the taxi outside, his colleague, Gill Burnside, was peering closely at her watch, but she smiled brightly and said, 'There's no hurry.' The taxi took off like a racing car.

By six-twenty Dyce Airport was packed with scruffy, hard-looking men wearing quilted jackets and toting duffel bags. None of them seemed to be over forty and most looked a great deal younger. They made me feel old and tired and inadequate; they made me wish I had at least had time for a cup of coffee. A pleasant Scots voice on the PA system announced departures for London, Amsterdam, Bergen, Sumburgh, Norwich. On the far side of the airfield, in front of the white beehive of the world's busiest heliport, lines of choppers, rotors whirring, were queuing for takeoff: Pumas and Sikorsky S-61s in the red, white and blue livery of the Bristow Company, huge white British Airways Chinooks. Between six and eight in the morning there seems to be more activity at Dyce Airport than anywhere else in Scotland.

The oil fields in the southern and central parts of the North Sea are within range of medium-sized helicopters, but there are only two ways of reaching those in the north: either direct by one of the new long-range helicopter services or by flying in a fixed-wing aircraft to Sumburgh Airport on the southern tip of Shetland, then on by Sikorsky S-61. That was to be our route on that Monday morning in the middle of March 1983. Our plane was an old Viscount with narrow seats and big oval windows, a relic from the days when flying was still thought to be glamorous and the view was part of the pleasure. We flew east over the granite terraces of Aberdeen until we were just off the coast, then swung slowly north. To the east, towards Scandinavia, the sun was clear above a blue-grey sea flecked with white. Below us, a strip of beach curved north towards Peterhead. Beyond the beach was an elaborate patchwork of fields – squares and triangles and polygons – the rich Grampian farmland where the famous Aberdeen Angus cattle are reared. Immediately ahead of us, luminous with sunlight, a great wall of cloud veered up like a mountain range. Within ten minutes it had swallowed us. The stewardesses brought around plastic cups of coffee, the NO SMOKING sign was switched off, and the world began to seem tolerable. When we emerged briefly from the clouds half an hour later the sea was a dim grey sheet and there was no sign of land.

Gill Burnside, the young woman from Shell, had dark hair, lively dark eyes, and, unlike the rest of us, an expensive suntan, since she had just returned from her honeymoon in the West Indies. She was also the only woman on board and the stewardesses did not register her presence. Each time they addressed the passengers they said 'Gentlemen,' and each time Gill's eyebrows contracted irritably. Finally, one of the stewardesses realised her mistake and came to apologise. She leaned confidentially over the seat and patted Gill's shoulder. 'We never get women on this flight,' she said. She seemed professionally embarrassed but privately suspicious.

Across the aisle from me was a squat, knotted young man

with a headful of ginger curls and the letters L-O-V-E tattooed on the first joints of the fingers of his left hand. According to the prison code, he should have had H-A-T-E on the right hand, but when he eventually brought his hands together to roll a cigarette I saw the name of a girlfriend: B-A-B-S. The paperback he was reading had on its cover a girl whose flowing black hair and flowing white dress were outlined against a stormy sky.

We droned along in unbroken cloud for another hour and emerged only when we were a couple of hundred feet above the waves, with the lights of Sumburgh airfield directly ahead of us. The Shetland archipelago is shaped like a dagger pointing down towards Britain, and the airport is at its southernmost tip. The building is modern, bright, larger than that of most British provincial cities, and almost totally deserted. It was planned in the mid-1970s to cope with the flood of technicians and construction men who passed through when the work in the East Shetland Basin was at its peak. But by 1978, when the bulldozers moved in to clear the site and extend the runways of the old airport, the first stage in the construction of the offshore installations was already complete and the world was in recession. By the time the glimmering £32 million terminal was opened in 1979 it was already redundant. Yet it was supposed to pay for itself. Since the terminal had been built with government money for the convenience of the oil industry, it seemed logical that the industry should be made to foot the bill. So the government instructed the airport operators, the Civil Aviation Authority, to recoup the cost through landing charges. The more the traffic fell, the higher the landing charges spiralled. By 1980, it was ten times as expensive per passenger to land a charter aircraft at Sumburgh as it was at Aberdeen. But it is not in the nature of international oil companies to take kindly to the role of captive clientele. They protested politely, then more strenuously, but when it became obvious that their complaints were being ignored they began to investigate the possibility of direct helicopter flights from Aberdeen to the northern fields. Shell, for instance, approached British Airways, who approached Boeing Vertol, who modified their big CH-47 Chinook military helicopters for civilian use. The Chinooks can ferry forty-four passengers direct from Aberdeen in airliner comfort – piped music and a steward to bring around the coffee – and then, if necessary, return to Aberdeen without refuelling, or divert to Bergen or even Paris. Half of Shell's crew changes to and from the northern fields are now by Chinook from Aberdeen, and within four years of being opened, Sumburgh Airport is obsolete.

The main concourse is as big as an indoor sports stadium, empty, echoing, white. We queued up for pink helicopter boarding passes, then waited in small glum groups, perking up only to attend to the voice over the PA system. The atmosphere was still early morning, liverish and surly. When our flight was finally called we queued again while customs officers sifted like gold prospectors through each piece of luggage. They were searching not just for the usual firearms and bombs but for drink and drugs. The offshore installations are dry, and all the instructions visitors receive insist, in varying degrees of intensity and usually in capital letters, 'NO ALCOHOL OF ANY DESCRIPTION' and 'NO DRUGS (except for prescribed medicines)'. The reasoning is simple and irrefutable: since life offshore is hard, dangerous, exhausting, and, above all, constricted, anything that might disturb the fragile balance of power and set men against one another is strictly forbidden. A person caught with a bottle is sent back immediately to the mainland – 'the beach' – and is thereafter banned from work in the North Sea. The companies take this regulation so seriously that no one is even allowed to begin the journey if he shows signs of intoxication. For example, in August 1978, there was so much activity in the Brent field that Shell hired a Swedish cruise ship, the *Stena Germanica* – known affectionately as the *Screaming Geranium* – to ferry workers out from Aberdeen. Although the journey took twelve hours, anyone who walked up the gangway drunk was sent back onshore to sober up.

The customs men were efficient and friendly. They smiled a good deal and tried to make conversation. But this only increased the general surliness. The baggage search is the passengers' farewell to freedom; it marks the beginning of a constrained and regulated life. The atmosphere reminded me of the special train back to boarding school when I was a child: a mixture of apprehension, resentment and gloom.

Beyond the customs desk was a small room with benches around three walls. Along the fourth wall was a counter behind which were hung rows of bright orange survival suits. The man in charge of them sized up each passenger critically. 'Small,' he said to me and heaved a suit across the counter. It was surprisingly heavy. I took off my shoes, struggled into it with difficulty, and put my shoes back on, while a bored official explained to us what to do in an emergency. The suit was stiff, rubber-lined, and equipped with a bewildering assortment of zips and flaps. The bottoms of the sleeves were turned back and the cumbersome rubber gloves attached to them were tucked into a flap at the shoulders. On the right shoulder, like an epaulette, was a strobe light, which is supposed to be stuck into the hood in the event of ditching, although I had run out of breath before I had managed to reach across and release it. I tied a lifebelt around my waist and kept the suit unzipped as low as possible to avoid suffocation. Outside the window a helicopter squatted on the tarmac, its rotors whirring angrily.

When the flight was finally called ten of us scurried out with our luggage, moving awkwardly in our unwieldy suits. The downdraught of the roaring blades seemed to be trying to crush us into the ground. The helicopter was a red, white and blue Sikorsky S-61, with the operator's name, Bristow, painted modestly in white on the red. The ground crew dumped our bags into a tail compartment, checked that our seat belts were fastened, slammed the cabin door, and scattered. The din of the engine and the rotors increased almost to the threshold of pain and the aircraft trembled slightly, as though preparing for a feat of strength. Then it suddenly rose straight up, like magic, like a creature set free from chains, hovered ten feet above the ground, dipped its nose as gracefully as a ballerina curtseying to her audience, swooped briefly forward, then up over a small headland to the misty sea. Patches of sunlight faded quickly as we climbed into the cloud.

Because I had never been in a helicopter, Captain John Hopson, Shell Expro's head of aircraft services at Aberdeen, had arranged for me to fly in the jump seat between the two pilots. Hopson is an ex-Navy flyer and looks the part: tall and athletic, with greying hair that makes him seem distinguished rather than old, and clothes as immaculate as a dress uniform – sharply tailored blue blazer, grey flannels with creases like swords, candy-striped shirt, white collar, club tie. He joined the Fleet Air Arm at seventeen, learned to fly fixed-wing aircraft, switched to helicopters, then joined Shell as a pilot in Brunei, the only place in all its worldwide operations where the company flies and maintains its own helicopters instead of contracting them from outside companies. In 1980 Shell brought Hopson back to Britain to supervise North Sea operations.

When I met him in Aberdeen I asked him if helicopters were much harder to fly than conventional aircraft, which is like asking a tennis pro if the backhand is harder than the forehand. He answered patiently, 'It's just a question of technique. Certainly, there are more things to concern yourself with when you're flying a chopper, more things going round. So learning is more difficult. If two pilots were going through initial training – one on a fixed-wing aircraft, the other on a chopper – and they compared notes, it would be reasonable to say that the helicopter was more difficult. But once you're flying professionally, there's no real extra difficulty. The basic difference is that a fixed-wing aircraft is inherently stable. With all the modern electronics – the automatic pilots and flight control systems – you can virtually fly a fixed-wing without touching the controls at all. You get it airborne, trim it out, take your hands off, and the thing will just go on flying until the engine runs out. If you take your hands off the controls in a Sikorsky or one of those little Bells, you crash and burn. You see, a helicopter is basically an unstable platform. So the electronic flight control systems and automatics are far more complex and expensive than those for a fixed-wing aircraft. That's why they have been so slow to arrive, although they are coming in now. There are a lot of modern helicopters that are pretty well hands-off from takeoff. Theoretically, at least. But with the older designs you have to fly them all the time. So it's more demanding physically and the speed at which the brain races is probably higher in a chopper than in a fixed-wing. But you get used to anything. Anyway, the difficulty is part of the satisfaction. I used to think I was very lucky to earn a salary doing something I enjoyed so much. Of course it's demanding, but for me it was also a very emotional job. The passenger may think that flying from A to B is utterly boring, but when you are at the controls the simplest things can stir you up: things like flying above cloud on a moonlit night or seeing the runway lights appear through the fog, just where they should be, at the bottom of an instrument-landing approach – seeing them there and knowing you're all set up beautifully. There's a tremendous satisfaction in that. If you've ever read Saint-Exupéry, you'll know what I mean. I suppose flying is a very selfish profession; you get magnificent feelings from it.'

I asked him how hard it had been to change from flying to office work. He answered, 'Of course you miss it when you stop. So you have to try to develop similar feelings about other things – such as bits of paper. But there's a challenge in doing any job properly. When I first came here I had great ideas of flying every other week and keeping my hand in. But when you come down to it, it's not possible to stay current as a professional once you stop flying as a way of life. Unless you fly every day, or so many hours a week, you're just playing at it. You lose confidence and after a while the desire to get airborne goes away. I found that even when I was a professional. In the periods when, for one reason or another, there's not much flying and the pilots aren't being used as they should be, you're quite content lounging around the crew room. Then when someone comes and says, "Go and fly," you think, Why should I? I'm very happy sitting here reading *Playboy*. Yet as soon as you start flying busily again, you know that the last thing you really want to do is sit around in the crew room and secretly you've just been waiting to get airborne. You see, once you're off that high emotional plane it's very difficult to bring yourself up there again, and once you're up there you don't want to get down.'

That high emotional plane seemed a long way from the bright, glass-walled office in the Tullos Industrial Estate in Aberdeen, and Captain Hopson laughed briefly, self-deprecatingly, as though to distance himself from youthful folly. It was a reaction I was to encounter often in the oil men I met, both onshore and off. A giant company like Shell, with an annual budget, technical resources, manpower, and a merchant fleet larger than those of many small nations, demands of its employees an allegiance akin to patriotism. They are required to change homes, towns, countries without argument, to switch jobs and work at the new one in a new place with the same unflagging intensity. They are also required to keep their pleasures and dissatisfactions to themselves. Ric Charlton, for example, at that time director of operations for the whole of Shell's North Sea operations, is a tall, restless Australian with a mottled complexion, uneven teeth, and the bridgeless boxer's nose and rangy physique of Jack Palance. He trained in Holland and since then has worked for Shell all over the world – in Nigeria, Kuwait, Trinidad, Brunei, as well as in Britain and his native Australia. He calls himself 'the classic oil field gypsy. Home is where I hang my hat. I had been seventeen years in the international oil business before I went back to work in Australia, my own country. And I probably felt more of a stranger there than I do here in Aberdeen. One fits in. And unless one does, one quits. I won't survive as an oil man unless I can move to Nigeria or Brunei and handle the minor aggravations you get in places like that and find satisfaction in the work. It's a process of self-selection.' Perhaps that is why so many of the people employed in the North Sea oil fields are ex-servicemen, used to obeying orders.

Both pilots of the Sikorsky from Sumburgh had learned to fly in the RAF before they joined Bristow. Because of the noise of the engine and the rotors we could communicate only through headphones and a microphone, like a black drinking straw, fixed in front of the mouth. I listened while they exchanged information from the duplicate displays of instruments set out before them, checking pitch, yaw, air speed, engine speed, altitude, and compass readings as we went up through the clouds to three thousand feet, then slowly descended until we were cruising in the sunlight at two thousand feet, just above a bumpy sea of cloud. When I had squeezed into the jump seat they introduced themselves, but only by their first names, as though they had too much on their minds to bother with unnecessary details. The chief pilot's name was Geoff. The second pilot was called Martin. Both of them had tired faces; Geoff's hair was grey. He told me he had been flying helicopters for eighteen years and was about to retire. 'Enough is enough,' he said. 'I'm going to buy myself a little shop somewhere. You know, a village store. The quiet life.' Over the din of the engines, I shouted into my microphone that I thought he would be bored. 'Yes, well,' he said, 'I don't exactly see myself sitting behind a counter all day.' Then he turned his attention back to the small yellow radar screen at the centre of the instrument console. The beam swung monotonously backwards and forwards, leaving a smear of buttery light.

A few minutes later, he said, 'Gliding. That's what I'd really like to do in my spare time.'

'Right,' said Martin. 'Me, too.'

'All that silence.'

'Right.'

Yet despite the noise and vibration, it was curiously peaceful up there above the clouds in the pale spring sunlight. We had left Shetland behind us and the neighbouring island of Unst, which has the northernmost airstrip and radio station in the British Isles, and now we were heading northeast, towards the middle of the North Sea and the sixty-first parallel, the same latitude as Anchorage.

While Martin watched the instruments, Geoff talked in a desultory way about flying helicopters. 'Nothing to it,' he said. 'Provided you pay attention.' He told me the only time you use maximum torque is on liftoff and landing. 'All you need is a bit of wind,' he said laconically. 'It's easier to land on an installation in a fifty-mile-an-hour wind than in dead calm. No wind, no lift, see?'

'If you say so,' I answered.

Above the radar screen was a small illuminated chart on which our progress was tracked automatically. We came in through the 'gateway' of the radio beacons at the southwest edge of the northern oil fields, but below us there was only cloud. Our destination was *Treasure Finder* – a mobile accommodation rig or 'flotel' – moored at that time alongside Brent Delta, one of the four fixed platforms (the others are Alpha, Bravo, and Charlie) that make up the Brent field. (All of Shell's North Sea oil fields are named after seabirds: the others are Auk, Cormorant, Dunlin, Eider, Fulmar, Gannet, and Tern.) As we approached, we lost height steadily, searching for clear air, but the cloud merged into fog and showed no sign of lifting. The pinhead of light on the chart showed we were getting very close. Geoff had throttled back the engines and he moved the aircraft forwards and downwards cautiously, as though groping down a darkened staircase. At two hundred feet the fog cleared for a moment, and there below us was the lumpy sea. Half a minute later the fog shut in again. Geoff was in urgent, muttered communication with Brent Log, the flight control. He eased the drumming aircraft down very carefully. When the sea reappeared it was so close that I felt I could lean out and touch the tops of the waves. The pilots talked to each other encouragingly, peering at the chart, then out through the window where the wipers thumped busily to and fro. Suddenly, Brent Delta loomed out of the fog on our right. It was much higher than we were and seemed enormous, with the waves beating around its huge concrete legs. Far above, from the top of its gas-flare tower, a great flowing rag of flame poured out into the fog. Martin said, 'Thank Christ.' Geoff merely grunted. The radio chatter increased to a grand, incomprehensible finale, like a flock of bickering seabirds. We curved upwards three hundred feet, circled Brent Delta, giving the gas flare a wide berth, then dropped gently down to the helideck of *Treasure Finder*. We were the last flight to make it that morning.

2

The first thing you do when you arrive on any offshore installation is arrange how to get off it again in an emergency. Lugging your bag, you scurry down flights of metal steps, following signs to 'Admin', a small, crowded room with a high counter where you hand your pink helicopter boarding pass – which doubles as a POB (Passenger On Board) voucher – to the Administration Office clerk. The clerk, who invariably seems to be the most even-tempered man on board, gives you in return a strip of coloured cardboard with your name on it and tells you the number of your cabin and your lifeboat station. You struggle out of your survival suit, go to your lifeboat station, and slide the cardboard strip into a slot on a wooden board. Should an alarm sound, everyone on board reports immediately to his lifeboat station, removes his name from the board, and puts on his life jacket. When all the boards are clear everyone has been safely mustered and the installation is ready to be evacuated. It is a simple and logical procedure, but not one that encourages optimism about life out in the North Sea.

Treasure Finder, the flotel for the Brent field, is owned and crewed by the Norwegian shipping company Wilhelm Wilhelmsen, was converted by them from an oil rig, and is contracted to Shell Expro. In construction, it is an Aker H-3 semisubmersible: a gigantic catamaran – a coffee table as big as a football field and four storeys high, mounted on two submarines. The accommodation is arranged in three decks: two- and four-berth cabins for five hundred men, a big mess hall, a coffee shop that is open round the clock, a cinema and two smaller TV rooms, a little gymnasium, separate rest and coffee rooms for the field's helicopter pilots (there are eight of them and they are all based on *Treasure Finder*), a corridor of offices, a radio room, a machinery room and various workshops and stores, and a glass-fronted box, like a foreshortened ship's bridge, where the Norwegian captain steers the vessel when it is under way. There is also a bonded store for duty-free goods – tobacco, after-shave, perfume for the wife – known as 'the bond'. Built on the top of all this are two helidecks and a hangar that can, at a pinch, house five helicopters and is painted on the outside in yellow and black stripes, like a wasp. On either side of this upper deck are a pair of massive cranes for unloading the supply boats.

In bad weather the journey from the helicopter to the Administration Office can be exciting. The helideck is covered with a rope mesh that provides something to grab on to should a sudden gust of wind blow you off your feet. Even in calm weather the noise is battering: the din of the chopper, the slosh of the waves around the huge legs of the adjacent installation, the roaring of power generators and the massed machinery as the oil is pumped and processed, the clank and thump from the drilling floor, water gushing from pipes high above the sea, the gas flare howling off into the murk. And over all this chaos on that first morning, every thirty seconds there came three long, melancholy blasts from the foghorn, the navigational signal for 'You are standing into danger.'

Below decks is calmer but never silent. The roar of generators becomes the discreet hum of air conditioning and there is a continual, subdued, hollow din of people moving around what is, however cunningly disguised, a steel structure. But the corridors are wide and rubber-carpeted and almost everyone wears slippers, since work boots are forbidden in the living quarters. In the cabins the metal doors and shelves are painted

to look like wood and most other surfaces are a yellowish buff that might also be wood. The cabins are cunningly designed to get the maximum use out of the minimum space. On the left of the door is a hanging place for the general-issue work clothes – orange overalls, round-toed leather knee boots, yellow hard hat emblazoned with the Shell emblem. Next to that are two long, padlocked clothes cupboards, a door to the shower-lavatory that is shared with the adjacent cabin, a recessed sink with a mirror above it. At the end of the cabin, beneath a square window with a close-up view of a metal wall, is a small table. On the wall to its right is a reading light, a thermostat, a shelf and two chairs, side by side, then two wide, comfortable bunks, one above the other, each with its own reading light and thick, brown-striped curtains; below the bunks are four long, deep drawers. In the ceiling is a fluorescent tube and a big ventilation slot. With the door shut the only sounds are the hum of the air conditioning and the far-off, intermittent foghorn.

I unpacked my bag and joined the two young people from Shell's Public Relations Department for a tour of the vessel. It turned out to be less a tour than a royal progress. With her dark hair, dark eyes, and bright face, Gill Burnside would be a remarkably attractive woman in any gathering. Out here in the middle of the North Sea, she brought the place to a standstill. It was like following Helen around the ramparts of Troy. At the door of the main changing room a man struggling wearily out of his overalls paused, crucified, arms half-in, half-out of his sleeves. Someone dropped his boots to the floor and whispered, 'Jesus.' Someone stuck his head out of his cabin, opened his mouth to call a friend, and was unable to get out a word. An older man with short, steel-grey hair took off his glasses, polished them vigorously on his sweater, put them back on, and shook his head. There were no bawdy comments, no wolf whistles, only a wave of incredulous silence, as though the men were overwhelmed by this dazzling reminder of a life elsewhere. When we three finally went into the mess for lunch Gill collected her food, marched ahead of us to a table in the farthest corner, and sat with her back firmly to the room. 'I couldna' eat if I had to face them,' she said.

In another place she would have been embarrassed or irritated; here she was simply disconcerted. The effect she had – the gawping, the stunned silence – was not lustful, nor even particularly personal; it was, instead, a mixture of wistfulness and fatigue. Fatigue because work on an offshore installation is unremitting. The regular basic shift is six o'clock to six o'clock, and it is worked with a minimum of breaks. 'It costs three hundred million pounds to build one of these things and seventy-five pounds a day to keep a man out here, not counting his wages and his flight from the beach,' an engineer told me. 'So you don't want to see it all come to a halt while six men take a tea break.' The cycle is monotonous and unforgiving:

up at five, eat, work, eat, work, eat. By the end of the minimum twelve-hour day – or night, depending on the shift – no one has the energy for anything more taxing than a game of billiards or a film before bed. The atmosphere in the rest area is friendly but drained. Twelve hours a day for fourteen straight days equals 168 hours' work – it is often nearer 200 – the equivalent of over four normal forty-hour weeks compressed into a fortnight.

As for the wistfulness: workers in the North Sea oil fields earn high wages but they pay for them with their private lives, and Gill Burnside's presence on board seemed to remind them of what they were missing. Depending on the firm that employs them, workers spend, at best, half their time away from their families. Shell employees are the luckiest; the company made a statistical survey that revealed that a high proportion of accidents occur after the workers have been in the field for ten days; as a result, Shell now rotates its men in one-week cycles. Most of the other oil companies and subcontractors work on a fortnightly turnaround – two weeks in the field, two weeks at home – although a few work three weeks offshore, one week onshore. Because of the complexities of saturation diving, the divers work for four weeks, then have four weeks off; but the divers, as in everything else, are special.

Intense work combined with an intense desire to get home makes men stir crazy. 'If you share a cabin with a man who clicks his finger joints while he reads in bed, you end up wanting to kill him,' a welder told me. 'Shell are right to do a week on, a week off. Two weeks out here is too long.'

The reaction to this off-on life varies according to the worker's background. For those who come to the oil industry from the armed services, the North Sea is a soft option: the food is better, the comfort greater, and they get home to see their families regularly instead of being away for months on end. One man told me that he had left the Merchant Navy for a job with Shell in order to be around while his children were growing up. No matter that he saw them only alternate weeks; he had previously counted himself lucky to be with them more than a few weeks a year.

Ex-servicemen form the nucleus of the North Sea establishment; they run the installations and fly the helicopters. The general workforce is different and fits less snugly into the scheme of things. Many of them have come offshore from Britain's declining industrial regions. The factories where they once worked have gone bust and they have lost their jobs through no fault of their own. So they come out to the North Sea grudgingly, as a last resort, because the work is there and pays well. Ken Murray, an ex-Merchant Navy sailor who is now offshore installation supervisor (OIS) of Brent Alpha, described what happens: 'When a man comes out here he's got to change his whole way of life and his whole way of thinking. He's probably been used to working shifts in, say, a steelworks. After work he goes to the pub with his mates, on Saturdays he goes to a football match, on Sundays he can forget about everything. But out here he's working twelve hours a day for fourteen days on end. No drink, no wife, and if he comes out to do a specialised job, probably no mates either. If he's part of the general workforce – a construction hand or a plater or a welder or a pipe fitter – then he's not part of a regular team like we have here on Brent Alpha, he's just one of five hundred blokes on *Treasure Finder*. So he's called out at five to eat his breakfast and catch the shuttle to work. Well, in winter it's not very pleasant getting on a helicopter in the dark at six in the morning, with *Treasure Finder* rolling around and wind blowing snow in your face. You're in an alien environment and a long way from home.'

It may be that an alien environment is just what some of the men are looking for. They choose to work offshore because their lives back on 'the beach' no longer make sense to them. The cold, the exhaustion, the wild weather, and the underlying sense of danger that goes with moving around continually in helicopters are a release, an objective corollary for their uneasiness. 'You'd be surprised,' I was told, 'how many men come out here to get away from their wives. You'd also be surprised how many marriages fail after the husbands go offshore.' An operations officer working for Bristow told me that at one point half the helicopter pilots on *Treasure Finder* were divorced and one was separated. He said, 'You count two weeks out of every four as lost time and try to pack four weeks of living into two weeks at home. So it's no place for the newly wed. The first year or two of marriage are hard enough wherever you are. Working out here could make them impossible. The young wife is left on her own to cope with everything: the babies, the house, the loneliness. She gets exhausted, then depressed, then she goes home to mother. On the other hand, if a marriage is really solid, it tends to prosper. If the kids are old enough to be at school and the wife has work and interests of her own, both she and her husband appreciate the time out to get on with their own lives. It makes them both independent and means they can enjoy each other's company more because they are not always living in one another's pockets. Even so, crises at home – the child breaks an arm or the roof springs a leak – nearly always occur just after the husband has gone offshore.'

Whatever the men's backgrounds, life on a North Sea oil installation is sufficiently constricting and unnatural that no one needed Gill Burnside to remind him of the world elsewhere. And Gill, who had merely come to look so as to be better informed about her work when she got back to her desk in Aberdeen, was finding the situation increasingly oppressive. We ate in silence, and when we finished she walked out of the mess with her eyes down and shoulders hunched defensively. The foghorn was still booming as we climbed the dank steel steps back to the upper deck. 'If it doesna' lift, we'll no get back tonight,' Gill said gloomily.

We went for coffee to the office of the OIS of *Treasure Finder*, Ian McKnight, the Shell employee who is ranked below the vessel's Norwegian captain but is ultimately responsible for all company decisions and activities on board. He has a big metal desk to match his rank, but otherwise the room is cramped and permanently busy. The security officer has his desk and files at the end of the office closest to the door. In the no man's land between the desks is *Treasure Finder*'s library: three shelves of a small cupboard between filing cabinets and a short-wave radio, and opposite a permanently replenished coffee machine. The library works on a one-for-one system: you have to leave a book before you can take one. 'It's all right if you like science fiction and cowboys,' said McKnight. 'But anything good gets stolen.'

We drank coffee, the foghorn boomed, the gloom thickened. When Gill asked, 'What happens if we canna' get off tonight?' the question hung ominously in the air; there are no single cabins for visitors on *Treasure Finder*. 'We'll sort something out,' McKnight said. McKnight is a bulky man made bulkier by a large sweater. He wears thick glasses and is a stickler for detail. He worked in the paper trade until the recession made him redundant, then joined Shell and came out to the North Sea in 1976 when the installations of the Brent field were under construction. He began as a critical-path-and-methods engineer and went on to do logistics for the engineering side of the project teams – logistics comprising everything from crew changes and matériel to planning. When the main flurry of construction was over he stayed on as OIS of *Treasure Finder*, a job that also requires a considerable talent for logistics since it involves accommodating and moving around the field a large and constantly changing population of workers, specialists, and visitors from Shell and its many subcontractors. A list of all the people who pass through the flotel each day, the firms they work for, the jobs they are doing, the installations they must be shifted to and from, and the places they will sleep is printed out daily on a telex sheet sixteen feet long that seems, to the outsider, as lengthy and impenetrable as the Dead Sea Scrolls. But despite McKnight's reassurances, Gill did not seem consoled.

I went down to my cabin and stretched out gratefully on one of the bunks. An hour later, I was woken by a small man with curly reddish hair. 'The name's Alex,' he said. 'Out here to check the copying machines.' He dropped his duffel bag on the floor and unpacked swiftly, with the air of someone who knew his way around. I asked him when he had arrived. 'Just a few minutes ago,' he answered. As though in confirmation, the cabin vibrated slightly and we heard the distant roar of a helicopter taking off. Gill Burnside and David Betts would get back at least to Sumburgh that night. 'Three M,' Alex said, when I asked him which company he worked for. 'There's three of us covering northeast Scotland. This, believe it or not, is part of our beat.' He unlaced his shoes. 'It's been a long day,' he said, 'and it's not over yet. I think I'll take a wee nap.' He placed his shoes neatly on the floor of his cupboard, climbed into his bunk, and pulled the curtains. He seemed to know that the secret of sharing a cabin with a stranger was to keep the place tidy and the conversation short.

At six o'clock, before going to eat, I went up on deck. The mist had thinned, although the foghorn still sounded its melancholy note. Brent Delta loomed over the flotel, lights blazing on half a dozen levels, machinery thumping and humming over the slosh of the waves around its legs. Its flickering gas flare illuminated the scene like a Hieronymus Bosch vision of hell. A mile or so away, Brent Charlie was lit up like a Christmas tree, the banner of gas above it blazing off into the dusk. Beyond it, Brent Bravo was a vague tangle of lights, and on the horizon was an isolated candle of flame from the Brent field's gas vent. Behind was another faint glimmer: the Dunlin platform. Smaller lights moved on the water: the supply ships that bring matériel out from the beach and the rescue boats that constantly circle each installation. But the gathering night and the fog reduced every detail to an Impressionist blur.

After dinner I began what I subsequently discovered was the regular evening ritual of everyone on board: migrating from recreation room to recreation room in search of a film worth watching (I finished up with a snow-and-ice James Bond). At ten o'clock I went back on deck for a breath of air before bed. The fog had lifted and the night was cold. The Brent platforms – Charlie, Bravo, Alpha – glittered in a straight line ahead, and in whichever direction I turned were other lights, less distinct and further off, rising out of the blackness. For the first time I began to understand the scope of this modern miracle: a network of little towns – built upwards instead of spread out – rising out of one of the most hostile seas on earth somewhere between a set of lost islands far off the northern tip of Britain and the harsh Norwegian coast.

A door opened behind me and a wiry little man came out, looked round, zipped his North Cape jacket up to his chin, and began to walk up and down the steel grating of the deck, moving his bent arms horizontally backwards and forwards and breathing deeply. After three or four turns, he stopped near the rail where I stood and nodded brusquely at the shimmering lights spread around us in the darkness. 'You know what that is?' he said and did not pause for an answer. 'That's what keeps the old country solvent.' He stretched his arms above his head and touched his toes six times. When he had finished, he said, 'That, mate, is the most expensive real estate in Britain.' Then he turned smartly on his heels and marched back inside.

3

In 1959, when the giant Groningen gas field was discovered on the north coast of the Netherlands, drilling for hydrocarbons in the sea was still in its infancy. The first offshore rig was used off Louisiana in 1947, and no one had experience of jack-up rigs and pipe laying at depths greater than two hundred feet. 'Around 1960 the North Sea for the oil men was like the moon for astronauts,' said George Williams, the geologist who headed the Shell operation for Shell/Esso's joint venture in North Sea exploration. The parallel between the North Sea and space has become one of the clichés of the industry. Shell recently ran a series of dramatic advertisements depicting science-fiction structures patrolled by strange spacecraft; the print below explained that this is what the oil installations and the pipe laying and submarine vessels would look like if the sea were removed. The message was that there is nothing to choose between the space programme and North Sea oil production in terms of both state-of-the-art technology and expense. Shell/Esso alone – and there are over forty other operators in the area – reckon that their share of the effort has cost them so far about £11 billion.* The Brent field itself was

* The word *billion* is used in this book in its American sense – that is, a thousand million. We often use *billion* to mean a million million, but the oil industry throughout the world follows the American style.

twice the cost of the Anglo-French Concorde project but, unlike Concorde, it coins money. Shell/Esso's average production for 1983 was 682,978 barrels per day at roughly $30 a barrel; in January 1984 the daily average had risen to 819,317 b.p.d. and it continued to rise through February to around 920,000 b.p.d. The total daily production figure for the UK sector is 2.4 million barrels – worth £630 a second, according to *The Times* – putting Britain in the first division of the world oil production league, along with the United States, the Soviet Union, Saudi Arabia, Mexico, and Iran. When the first British oil – from Hamilton Brothers' Argyll field in the central North Sea – came ashore in June 1975, the British national oil bill was £3.791 billion a year. Britain is now self-sufficient in oil and currently receives from its oil and gas fields an annual tax revenue of £12 billion a year. Despite its ailing and largely inefficient industries and its minuscule rate of economic growth, the country now has a regular balance-of-trade surplus because of oil.

Long before the discovery of the Groningen gas fields, geologists had surmised that there might be hydrocarbons in the deeper waters further north, where the gas-bearing formations of Yorkshire dip under the sea to reappear in the gas fields of northwest Europe. Groningen showed that the deposits under the sea might be large enough to be commercially profitable. But no drilling was possible until the governments of the various countries around the North Sea had ratified the Continental Shelf Convention, setting out which part of the sea belonged to whom. The Convention had been agreed in principle in 1958 but was not finally ratified until 1964. Although the UK, ironically, was the last to sign, it was one of the first to issue licences for exploration and production. By then, the oil companies, inspired by the Groningen bonanza, had already done a series of seismic surveys of the southern section of the North Sea and were ready to move. Fifty-three licences were issued in September 1964; the first exploration well was spudded in before the end of the year, and twelve months later the first offshore UK gas field, West Sole, had been discovered by BP – to be followed soon after by Shell/Esso's strike in the giant Leman gas field. By 1967 West Sole was producing and the conversion of British homes to natural gas had begun. Virtually all domestic gas in Britain now comes from the North Sea.

Despite the gas finds and a small oil discovery in Danish waters in 1967, not many oil men seriously believed in the possibility of commercial oil deposits under the North Sea. The two exceptions were Edwin van den Bark of Phillips Petroleum and George Williams of Shell. The gas fields were on the latitude of the Midlands, between the fifty-third and fifty-fourth parallels, but van den Bark was convinced that oil would be found further north, above the fifty-sixth parallel. His enthusiasm and persuasiveness were vindicated by two strikes in the Norwegian section: first, the small Cod field, then in 1969 the giant Ekofisk. That same year Amoco struck oil in British waters above the fifty-seventh parallel, and in 1970 two more fields had been discovered in the British sector east of Ekofisk: Phillips's small Josephine and BP's Forties, another giant. The following year, in the same area, Shell struck oil in the Auk field.

Meanwhile, George Williams, inspired by recent developments in the geological science of plate tectonics, had become convinced that the real bonanza was to be found much further north, around the sixty-first parallel. Williams has recently retired as director general of the United Kingdom Offshore Operators Association, which provides the companies involved in British offshore oil with a channel through which to communicate with the outside world – government, the media – and a forum for internal discussion of their problems. Williams is a vigorous, ruddy-faced man, now in his sixties and putting on weight around the middle. But even in the sedate Knightsbridge offices of UKOOA he moves as though holding in with difficulty a great deal of energy. His eyes are clear blue and very steady, his mouth is bitten up and screwed down on the left side, although it straightens when he smiles. He has a booming, barrack-room laugh. He comes from a farming family in Wiltshire and when he went up to Cambridge in 1936 he had no idea what he wanted to study. When his tutor suggested geology, he told me, 'I thought, Well, I've been brought up on a farm; that sounds interesting. Then when I began to think of what I was going to do with my life, I saw geology as a means of seeing the world. Just before I finished my degree, BP offered me a job, but I wasn't attracted by it because at that time they were only in the Middle East. So I went over to The Hague and was interviewed by Shell, who offered me a post, provided I got my degree. Then I thought that perhaps I ought to test the water with another company as well, so I wrote to Burmah Oil. Believe it or not, I got a reply back in forty-eight hours, saying, "Dear Mr Williams, I was most surprised to get your letter. I thought you were close to finalising arrangements with Shell." I thought, Good God, they're the Secret Service; I'd better not try anyone else. So I quickly signed the Shell papers. I got my degree in 1939 and delayed joining Shell, with their blessing, to do a few months' field work – because I'd won a prize of some sort. I was due to go to The Hague on the fifth of September. Well, war was declared on Sunday, September the third, and at nine o'clock precisely on Monday the fourth, at my home in Wiltshire, I got a call from Shell's chief personnel director, no less. The conversation went like this: "Williams, I presume you've heard that war's been declared. Knowing that you'll want to join the services straight away, the purpose of this call is to tell you that we've cancelled arrangements for your passage to The Hague tomorrow and to assure you that there's a job for you in Shell as soon as you've won the war." ' (Big laugh.) 'I was a bit bloody annoyed.'

Williams joined the RAF, became a pilot, and had what is called 'a good war' – so good that when it ended he was tempted to stay in the services. It took him, he said, a long time to decide, but when he eventually went to Shell as an exploration geologist he got all the travel he had originally bargained for as an undergraduate: prospecting in Somaliland, Indonesia, and Libya, broken by spells in The Hague, the United States, and back home. In 1954 he became exploration manager for Shell's joint venture with BP in Nigeria. Despite the fact that the companies had been prospecting there, on and off, since 1935 without success, Williams became convinced there was oil to be found. Not so The Hague office. When he returned to Europe in 1955 with his proposed programme for the following year, he was met, he said, 'with a great lack of enthusiasm'. But that was precisely the kind of challenge he most enjoyed. He returned to Nigeria having wheedled the money to drill three more wells. Before the year was out, they had made their first major strike. He then moved on to Brunei, where Shell had just begun to explore the offshore waters. 'When I arrived they'd drilled three wells, all dry. We carried on and drilled another eighteen dry holes. Then I came home . . .' (Big laugh.) '. . . and presented to the board my case for continuing and had the same problem as I'd had with Nigeria.' (Another big laugh.) 'Anyway, I managed to persuade them. We drilled one more dry hole and in the twenty-third well we struck oil.' The Sultan of Brunei, Sir Muda Hassanal Bolkiah, thanks to his oil revenues, is now one of the richest men in the world, with a palace that covers three hundred acres.

Williams's track record in Nigeria and Brunei stood him in good stead when what he calls 'the third sticky point in my career' arrived: the decision to drill for oil in the northern North Sea, around the sixty-first parallel. The expense was enormous, the technology undeveloped, the Shell board doubtful. Added to that, one senior Shell geologist went on record as saying he could find no evidence for source rocks or good reservoirs in the area. 'But I was an enthusiast. I think everything I've ever been given to do in life I've been enthusiastic about – perhaps because they've been good things to do. In the North Sea I was fired by the whole project from the start and then became more and more certain that all the signs were right. Of course, there's an element of luck in oil exploration. Not gambler's luck, simply the luck of being in the right place at the right time and then taking advantage of it. The odds are usually against you, so if you're going to be successful in this business, you must try to reduce those odds to a minimum. And if they are still too great you must say so and get the hell out. But if you think there's a chance of success, you've really got to hammer the table and convince other people.'

Williams hammered the table and eventually the Shell board was convinced. In 1970, when the British government opened the third round of exploration licensing, the company's geophysicists had analysed extensive seismic surveys of the East Shetland Basin and discovered what Williams called 'very marked geological unconformities at varying depths between six and twelve thousand feet which we hoped would be an unconformity in the Cretaceous-aged rocks. In other words, below the unconformity we would run into lower Cretaceous and Jurassic rocks. If that was the case, we were going to have a suite of rocks down there that would clearly be very good from an oil or gas point of view. However, the unconformity could have marked the boundary between upper Cretaceous rocks and much older rocks – Devonian, for instance – which would have meant little hope for hydrocarbons. There was no way we could tell until we drilled.'

Hydrocarbons – oil, gas, tar – are derived from the slow decomposition of organic matter – microscopic animals and plants – trapped many millions of years ago in accumulating muds. Their decomposition was encouraged by ever-increasing pressure and temperature as the muds were gradually buried under further sediments. Over the millions of years hydrocarbons take to form, the increasing pressure squeezes them out of the gradually consolidating muds into adjacent, more porous sediments, such as sandstones, limestones, or chalks. Contrary to popular belief, hydrocarbons do not gather in subterranean caves and then gush to the surface when a drill breaks through the impermeable cap rock holding them in place. Instead, they are usually found in the minute spaces between grains of consolidated reservoir rocks – sandstones and limestones – to which they have migrated from the original source rocks – the fine silts or shales from which the hydrocarbons have been expelled. Reservoir rocks, which are porous and permeable, consist of a network of interconnected, water-filled pore spaces through which the gas and oil gradually permeate upwards. A three-dimensional configuration of reservoir rocks, surmounted by a cap rock that prevents the hydrocarbons migrating or escaping, is called a trap. These occur only in layers of sedimentary rock – as distinct from the other two groups of the earth's rock, the igneous and metamorphic – that were laid down when the earth's crust was warped downwards, usually under seas, sometimes under lakes and estuaries, to form sedimentary basins.

Source rocks, reservoir rocks, cap rocks, a trap, and a sequence of geological events that produces these four requirements at the right time and in the right place are the basic geological pieces in the jigsaw puzzle of oil exploration: Is there any indication of oil? Is it mature? Where are the source rocks? Where could it have been generated? Where might it have migrated to? Geologists obtain the preliminary answers to these questions from seismic surveys: acoustic shock waves are passed through the strata and are reflected back from the various layers of rock; their times of travel are measured by recording instruments, and from the readings pictures are produced of the underlying formations; from these it is possible to make an educated guess as to the probable presence of oil or gas. If the signs are right, you are in what oil men call 'an exploration play'. But the only way to be certain is to drill.

In the early days, before petroleum engineers knew how to control the flow from a new well, the classic indication of a strike was the gusher, a great unstoppable fountain of oil that roared into the air and flooded the surrounding countryside. In *The Wildcatters*, a history of the early American oil explorers, Samuel W. Tait describes what happened in 1910 when the Lake View well, on the east side of the Coalinga field in California, blew out:

> With a roar of gas that could be heard miles away came a cloud of oil whose particles carried for two-and-a-half miles . . . Dusty [Charles L. Woods, the foreman] paid high wages to college boys, convicts and anybody else who cared to help pile up sand bags around 75,000 barrel sumps. Outside the sumps, which were filled as fast as they were completed, he walled up a sixteen-acre area which he named the 'Cornfield'. One of California's unprecedented earthquakes cracked open three sumps and let oil into this. To prevent the late summer floods from taking out the confined oil, he led six hundred men back to the hills and dammed up the mouths of canyons with walls twenty feet high and fifty feet wide. Nine-tenths of the storage he built was needed before the big well died.
>
> Meanwhile persons were composing a variety of descriptive passages about the Lake View. To one onlooker it was utterly unearthly. 'It's hell,' he said, 'literally . . . hell. It roars like hell. It mounts, surges and sweeps like hell. It terrifies like hell. It is uncontrollable as hell. It is black and hot as hell.' Charley Woods' description was . . . more pointed. He said the earth had cut an artery.

Today the signs are outwardly a good deal less dramatic. While an exploration hole is being drilled, a chemical compound called mud is pumped down the drill pipe, emerges from holes in the centre of the drilling bit, then circulates back up the hole to the surface. The mud has several functions: it lubricates and cools the bit; it lines the walls of the hole, preventing them from caving in and making it easier to retrieve the bit; it provides weight to balance the formation pressure encountered during drilling. But most important, on its journey back to the surface the mud flushes out the chippings made by the turning bit. Back on the platform, the chippings are filtered out of the mud, so that it can be used again, and are then analysed for information about the strata penetrated. Sometimes the mud also brings up traces of gas or oil, the first hints of a discovery. The next stage is to run an electric log down the well on a cable in order to measure the resistivity or conductivity of the rocks. (Since oil is an insulator and salt water a conductor, the resistivity of oil-filled rocks is high.) There are now other, more sophisticated logs: nuclear logs that bombard the formation with gamma rays, sonic logs that measure the velocity of sound through it.

George Williams described to me what happened in 1971 when Shell drilled its first exploration well in what subsequently became the Brent field: 'We had got to around seven thousand feet and were very much on edge to know what would happen when we went through the unconformity. We went on down and, as we had hoped, somewhere between seven and ten thousand feet we went through into an absolutely marvellous suite of Cretaceous and Jurassic rocks, full of oil. The first indication is your cuttings, but they can be deceptive. It's the logs that show the character and contents of the rocks, and when we saw them we realised we had a wonderful reservoir. But that presented us with a problem. The government had announced the fourth round of licensing. By that time we had found other attractive structures in the vicinity, one of which was block 211/21, which the government was putting up for auction in a few months' time. Now, because we are a highly competitive industry, we felt it was terribly important that the wide world should not know we'd found oil close by in Brent. So we decided not to carry out a full production test, which would have meant producing the oil and gas and flaring it off – something you could never hide. Instead, we did a Schlumberger formation test to determine the content of the rock. That means you put in a great long sonde – a sonde being an electronic logging tool that gathers data and sends it back to the surface. This particular sonde has charges up and down its length. When you fire them by remote control each one shoots a little cone into the rock. The cones then seal themselves and are tied automatically with a piece of wire. When you pull the sonde back up, these little sealed capsules come with it, each containing a sample of the rock. You can then analyse what is in the pores. That was when we knew that our greatest hopes had been realised.'

With the government auction imminent, it was vital to keep the results hidden, even from the drillers on the platform. According to Ric Charlton, apart from senior members of the Shell and Esso boards who had to decide what to bid, only four oil men knew the potentialities of this first well in the northern North Sea: George Williams, on the spot; 'one fellow in The Hague'; Jock Munroe, the base representative in charge of drilling, who flew with the samples from the platform to Aberdeen; and Charlton himself, who was in Lowestoft on the Norfolk coast opposite Shell's gas fields: 'I was petroleum engineering manager at the time, one of the people responsible for the evaluation of the wells. I'll never forget seeing those logs and samples. Eight hundred feet of oil-bearing sands! Fantastic! It was like breaking the bank at Monte Carlo. But we kept that well as tight as the proverbial duck's backside. We drilled it and nobody knew we had made a major discovery. Then we sat on it for nine months.' Jock Munroe later confessed that, although he had managed to keep the secret from his wife, he did once mutter it to his dog.

Shell already held a licence for the Brent field, but the government had decided that the next round of licensing, which included the nearby block 211/21, would, for the first time, be an auction by sealed cash bids. 'Remember, this was the very first well in the really northern North Sea,' George Williams said. 'And only we knew for sure that it was an oil province with prolific production possibilities. If the others had got wind of it, fantastic amounts would have been bid on block 211/21. So we kept our secret. But then, with Esso, we had to decide how much we would bid. There are various techniques for working out the sums, but the key factor was we were absolutely determined to get that block. We knew more than anyone else and could not afford not to take advantage of that. So after long deliberation we decided to bid twenty-one million pounds, or just a little more.' When the sealed bids were opened it turned out that the next highest bid was £8 million, which gave the rival companies a good deal of malicious pleasure. 'So what?' said Ric Charlton. 'So we left thirteen million on the table and everyone said, "Ha, ha." But I still think it was a good figure. If you are going to make a pre-emptive bid, you might as well make it pre-emptive. Think what we've got in that block: the bulk of Cormorant – a difficult field but nevertheless more than a half billion barrels of oil in total. What the hell is twenty-one million pounds?'

In the spring of 1972, when the fourth round of licensing was out of the way and the lid was off their secret, Shell drilled a second Brent well, which went through the main upper reservoir, and discovered beneath it a second reservoir, equally large. At that time, John Heaney, until recently both managing director of Saxon Oil and a fruit farmer in Essex, became technical director of Shell Expro in the northern North Sea. 'My job was to develop the field, to spend the money,' he said. 'Once we knew the shape and size of the field, we set up computer models to work out the economics of developing it. The key thing to remember is that oil at that time was two dollars and fifty cents, becoming three dollars, a barrel. We justified the development of Brent at that price. Within a couple of years, in the wake of the Yom Kippur War, OPEC got its act together. The price of oil is now ten times what it was when we originally did our sums.' As Ric Charlton said, 'What the hell is twenty-one million pounds?'

Before the discovery of the Brent field, George Williams had made a bet with a rival geologist in BP: if there was oil in the northern North Sea, said the man from BP, he would buy Williams the best and most expensive dinner in London and would also eat his hat. After the fourth round of licensing, he bought the meal but Williams let him off the hat.

4

The North Sea workforce moves around the fields only by helicopter. From six o'clock onwards, *Treasure Finder* shakes faintly but continuously as the choppers come and go, the roar of their engines a moving bass accompaniment to every activity. The flights are divided into two categories: shuttles and buses. The shuttles take workers to their jobs during the rush hours, which start at six in the morning and six in the evening and last a couple of hours. In the quieter time between, the buses buzz around the field, picking up and dropping off passengers, mail, and freight. The four helicopters for Shell's northern installations are based on *Treasure Finder*. Before takeoff each pilot is handed a slip of paper with his flight plan set out like the clues of a children's treasure hunt. His instructions for a typical bus run around the field, calling at the four Brent installations (Alpha, Bravo, Charlie, Delta), Cormorant A, North Cormorant, Dunlin, and the drilling platform Stadrill, look like this:

```
4    F  1 mb 1F  3                     2f     1         F
TF – D – C – A – B – N.COR – S.DRILL – COR – DUN – TF – A – TF
                 F     3       3mb      2     2    2f       F
```

The code is simple: numerals indicate the number of passengers; F means big freight, f small freight, mb mailbag; anything written above the name of the installation means pick up; anything written below it means drop off. All this coming and going, collecting and delivering, is worked out at Brent Log, the aircraft control centre on Cormorant A. Brent Log is run by Shell but controls flights throughout the whole East Shetland Basin, an area with a radius of roughly forty miles, encompassing BP's Magnus in the north, Chevron's Ninian in the south, and Union's Heather in the west. Brent Log coordinates the Bell 212s that shift men and matériel from installation to installation and it also controls the crew-change aircraft – the Chinooks and Sikorsky S-61s – shuttling to and fro from the beach. In 1978, when the fields were being built up and the platforms brought into operation, Brent Log handled a hectic peak of twenty-two thousand aircraft movements in one particular month. This made the tiny office in the middle of nowhere busier than any other air traffic control centre in Europe, except for Heathrow. In early 1983, when I was there, most of the building was finished and the platforms were pumping oil, and Brent Log had settled down to a mere twelve thousand movements a month – between four and five hundred a day – which put it at about the level of Gatwick, Britain's second busiest airport. Some of these movements are extremely short – the horizontal distance from the helipad on *Treasure Finder* to that on Brent Delta was a couple of hundred yards – even so, each has to be slotted into an overall flight pattern and monitored coming and going.

But no matter how meticulous the planning, a helicopter is, as the pilots say, 'an unstable platform' and occasionally – very occasionally considering the number of air movements – accidents happen. Shortly before I went out to Brent, a Sikorsky S-61 on a crew-change run from one of the more southerly fields went down off Peterhead, north of Aberdeen; all seventeen passengers, as well as the Sikorsky, were rescued unharmed. In July, when I was in the field again, another crew-change helicopter crashed coming into Aberdeen's Dyce Airport. The previous year, in September 1982, an emergency helicopter, going to rescue an injured diver, went into the sea with the air crew and medics, killing all six. No one talks much about these incidents, but in a place where helicopters stop being exotic machines used by tycoons and soldiers and become, literally, buses, used as regularly and casually as a city worker uses an ordinary bus, every accident builds up tension, adding an extra subliminal anxiety to the general strain created by the unnatural isolation of life offshore, the exhaustion of the work, the hostility of the stormy North Sea. 'Every incident digs into your brain,' I was told. 'It's nothing dramatic, but the bubble of confidence tends to collapse a little every time something goes wrong. So there's always a bit of nervousness, a feeling of apprehension, about getting on the damn things.'

Not so for the casual visitor. That first morning in the field, buzzing above a choppy sea from one clanking platform to another, like a bee at a flower show, all I felt was exhilaration. It was like being let in on some marvellous secret of human ingenuity and audacity. As we flew towards Brent Alpha, the most southerly of the four Brent platforms, the excitement of flying, of being there, of participating, in however tenuous a way, in the essential business of the world, combined with the sheer size and complexity and improbability of the installations scattered across the water, made me wonder why I had had to wait forty years to do things I had yearned to do as a schoolboy.

The offshore installation manager of Brent Alpha was Iain Blair, a former Royal Navy engineer. Blair is soft-voiced and very short. He has innocent blue eyes in a lugubrious bloodhound's face, a thick neck, sloping shoulders, and forearms like Popeye's. Although he is in his forties he still plays rugger and squash back on the beach and keeps fit offshore by running up and down the installation's innumerable stairs. Each morning, like all the other OIMs, Blair meets the five men responsible to him in the running of the platform: the offshore installation supervisor, in charge of the platform's general services – deck work, catering, drainage, helicopter movements; the production supervisor, whose job it is to keep the oil and gas flowing; the platform engineer, who maintains the production machinery ('The PS breaks it, the PE mends it,' said Blair); the Shell 'toolpusher', directly responsible to Shell for drilling operations, although the drilling itself is subcontracted to a German firm, Deutag; and the construction engineer, who oversees any alterations or additions to the platform. Each of the five reports on the state of his particular province and the work to be done during the next twenty-four hours; then together they discuss problems. The Shell toolpusher, Franz van Haaften, a blond Dutchman with a neatly trimmed beard, made his report in the fewest possible words, then sat in silence while the others had their say.

Dave Swankie, the construction engineer, had lank grey hair, grey stubble on his chin, a witty slit of a mouth, and a broad Scots accent. He had assumed the deadpan Buster Keaton demeanour of a man permanently oppressed by unruly machinery. In a funereal voice he announced, 'The air compressor has been moved to its final resting place.' (A two-beat pause.) 'I hope.'

While van Haaften watched impassively, the others began a heated discussion of the turbines that supply the installation with power. There are two of them – developed from the Rolls-Royce Avon jet engine – and one had broken down and the generator had to be shifted to get at it. The problem was that the generator weighed eleven tons and none of the girders around it was designed to bear loads greater than three and a half tons. They decided that the solution was not to lift the brute vertically but to slide it horizontally. This meant cutting away a steel staircase. 'No trouble,' said Swankie, po-faced. 'Just so long as they keep their acetylene torches away from the gas lines round the stairs.' This provoked a further flurry of anxiety and a lipless grin from Swankie.

'One last thing,' said the OIS, Ken Murray, who has dark floppy hair and a pleasant smile. Van Haaften glanced at his watch and began to fidget. 'There's a rumour that the Deutag roughnecks are bringing in porn videos from Hamburg and selling them to our chaps.'

'Nothing wrong with that,' answered Bob Paisley, the production supervisor, a young man with an incongruous flowered shirt and long hair. 'Buying a video is up to the individual.'

'Not when it's sold through the bond, it isn't,' said Murray grimly.

'Look into it,' Blair told him. 'Right, gentlemen, if that's everything for this morning . . .'

Outside a sharp wind was blowing and the sea glittered. I panted after Blair up a maze of staircases to the helideck. He swept the horizon in a proprietorial way and said, 'Nice view, isn't it?' There were installations in every direction I looked. I turned slowly, counting, while Blair reeled off their names: the three platforms of Chevron's Ninian field; Union's Heather; Shell's Cormorant A; Amoco's NW Hutton; Shell's Cormorant North with Dunlin in the background, and in front of them, quite near, Spar, a tanker-loading buoy and oil storage facility that looks like a gigantic, top-heavy pillar box; Britoil's Thistle; Conoco's Murchison; the three other Brent platforms and *Treasure Finder* in line ahead and close to us; two installations of the Norwegian Statfjord field; a Norwegian tanker-loading buoy; the Brent remote gas flare, like a solitary, admonishing, burning finger; two Norwegian mobile drilling rigs. Twenty-two constructions, a whole industrial province, rising out of the icy sea. The remote flare, which burns off the gas that can't be coped with by the pipeline to St Fergus, on the coast north of Aberdeen, blazed in the sunlight a mile or two away. Blair nodded at it and said, 'One second of that would heat your house for years.'

A small bird was perched on the edge of the helideck, wagging its tail feathers. I said, 'What's it doing here? It looks like a swift.' Blair shrugged. 'They get blown out here,' he said. 'Pigeons, crows, owls. I've even seen a merlin. Not to mention all the usual seabirds. Myself, I like 'em. They add a touch of colour, a little extra something to the routine. We had a man out here once, name of Crow, employed by a firm called Sparrow, found a quail on the helideck. It must have been carried from the beach on the undercarriage of a chopper.'

A small bluish-grey dot rose into the air above Cormorant A and moved slowly towards us, gradually evolving the bug eyes and angry snarl of a Bell 212. We moved to a little platform just below the helideck where two passengers, a wooden box, and a mailbag were waiting. While we watched the helicopter approach, Blair said, 'A couple of years ago when I was on Fulmar, three pigeons landed so exhausted they couldn't move. Took pity on the poor little blighters, crated 'em up, and sent 'em back to the beach on the next chopper. One of 'em was ringed, so the chaps at Tullos in Aberdeen popped 'em on a train – no expense spared – and sent 'em back to their owner, somewhere south of London. The next thing we know, the fellow's written a furious letter to Head Office, saying we'd ruined his chances in a pigeon race to Norway and why didn't we mind our own bloody business?'

The helicopter landed as gently as if the deck were lined with porcelain. There was a flurry of activity as a misshapen piece of equipment was unloaded from the tail compartment and dragged across to where we stood watching. The helideck crew, headphones under their hard hats, lugged the wooden box and the mailbag across to the chopper, signalled to the waiting passengers, bundled them in, secured the doors, and scurried back to cover. The Bell rose, hovered, dipped its nose, and swooped away, leaving us in blessed silence.

On the drilling floor, beneath the towering latticework spire of the derrick, roughnecks were busy finishing a 'round trip' – pulling the entire drill string back up the hole in order to change the bit. The pipe is usually five inches in diameter – although it can vary between three and a half and six inches – and comes in lengths of thirty feet. On a round trip it is pulled up three sections at a time by a gigantic hook and swivel hanging from strands of a continuous wire rope that runs up to a crown block at the peak of the derrick, over a pulley and back down to an even more gigantic drum on the drilling floor. The roughnecks – assistants to the driller – working rhythmically and with great precision, clamp multihinged tongs around the base of each three-length section of pipe that emerges. To prevent the rest of the pipe from dropping back down the hole, they hammer into the space between its protruding end and the rotary table on the drilling floor a linked ring of jointed and tapered narrow wedges with steel octopus-sucker pads on their inner surfaces. While two roughnecks heave and strain to keep the tongs still, the rotary table spins and the two sections of the pipe unscrew. The bared threads are oily but clean and surprisingly delicate. Everything else in the area is Brobdingnagian in size and covered with pungent, greasy drilling mud that coats the overalls and gloves and boots of the roughnecks, making them look like Roman wrestlers as they grapple with the base of the dangling stand of pipe and balance it carefully across the floor. Although the steel of the pipe is half an inch thick, it whips and flexes as though alive. High up in the derrick a man stands on a little platform, leans out over the interior space like a gargoyle on a church, and lassoes the top of the pipe. The clamps from which it is suspended are released and he stacks it dexterously in place with the rest of the withdrawn pipes. They lean against the side of the derrick like sticks of rusty candy in a jar.

Nine thousand feet below the icy sea the temperature in the oil reservoir is around two hundred degrees Fahrenheit, and even after the pipe's long journey to the surface it is still hot. It steams gently as the roughnecks hose the mud from it while it is withdrawn. Steam also rises from the four vent holes in the rotary table. It seems strange, among all this high technology, to find that perhaps there was some truth in the medieval theories of hell. As each new section is withdrawn, one of the German roughnecks, who seems to be doing most of the heaviest work, spits into the steaming hole like a dog signing a lamp post.

The last sections of the pipe from the deepest part of the well are drill collars, ponderous cylinders, much thicker than the drill string, that add weight to the bit and help keep the drill string in tension. They were interrupted every sixty or ninety feet by stabilisers, elegantly machined cylinders with wide bands of dark rifling. As each stabiliser is removed, the inner threads at its top and base are meticulously hosed down; then each end is capped with a heavy metal hat surmounted by a hemispherical ring. The new drilling bit waits on the far side of the derrick floor, pristinely clean and painted silver. It looks like a great clenched fist with cogged teeth where the

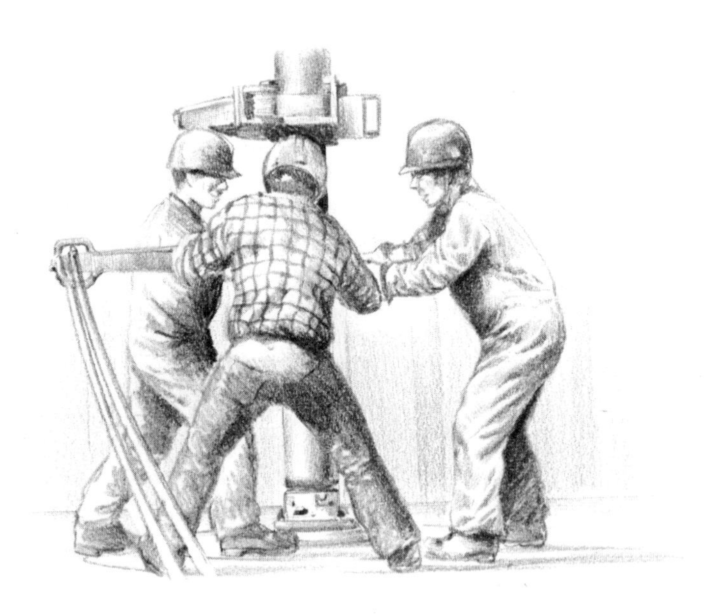

bent fingers should be. On its top, it, too, wears a shabby cap to protect the threads.

At the opposite side of the floor is a control cabin with grimy windows where a big taciturn German pushed levers and eyed a battery of flickering dials. It looked to me more complicated than the controls of the helicopter I had flown in that morning, but Blair was unimpressed: 'The popular belief back on the beach is that the whole of the North Sea oil industry is on the frontiers of technology. In some areas that's true, of course. But from an engineering point of view, a large part of it is downright agricultural. For example, if the bit gets stuck – and it happens all too often – they have to jar the thing out – jolt it back, like hammering in reverse. When they do that the whole platform shakes, even though the bit itself may be two miles down or more. Of course, when you look at the problems of putting a platform in the North Sea the actual construction of the things is quite clever. We're one of the smaller installations and even we are seven hundred and fifty feet tall: four hundred and fifty feet in the water, then another three hundred feet to the top of the derrick. Some of the big modern platforms with vertical gas flares are a thousand feet high from the tip of the flare to the seabed. That's taller than the Eiffel Tower. So you have this huge construction that has to contend with wind and water. They're designed to withstand what's laughingly called "the hundred-year storm" – hundred-foot waves and winds gusting up to one hundred and sixty miles an hour. So the design of the platforms, building the brutes, floating 'em out and then positioning 'em by satellite – all that is really quite clever. But once you get down to the basic services, like the power generation and distribution systems, the hotel services and the rest, there's nothing particularly clever about any of that. In fact, the power systems in a naval warship tend to be more sophisticated.' He eyed the rows of dials in the control cabin grudgingly: 'Occasionally, of course, we do buy some expensive bits of kit. But that's the name of the game up here. A development well, like the one we're drilling now, costs about three million pounds. An exploration well will run you twice that.'

(As a former officer in the Royal Navy – 'the silent service' – Blair is particularly adept at understatement, a British art that generates a language, as well as a style, all its own. 'Quite clever' is one of its key phrases and is interpreted most easily by a few facts: in 1981 Mobil's 890-foot-tall Statfjord B platform became the largest structure ever moved across the face of the earth by man; it weighed 785,000 tons and cost an estimated $1.8 billion. Its companion, Statfjord C, weighs in at 697,000 tons and Chevron's Ninian Central at 636,000 tons. In the April 1982 *Scientific American*, Fred S. Ellers wrote: 'To help build such immense structures semisubmersible derrick barges have been developed that are able to lift 5,000 tons, the weight of two World War II destroyers, nearly 200 feet above the water and to do so in the open ocean.')

Out on the drilling floor, the tempo and precision of the work had risen as the round trip came to its end. The stabilisers nearest the bit were as massive and carefully machined as prize exhibits in a museum of nineteenth-century engineering masterpieces. The last one to come up had a small brass plate let into its gleaming surface on which was engraved, like a hallmark on silver, '7-REG', a series of numbers, then again, '7-REG'. The final cylinder before the bit contained the Teleco tool, a relatively new instrument developed to measure the position of the bit and the angle of the drill hole. It uses the drilling mud that is pumped down the centre of the drill string to drive a small turbine that powers the tool itself. The mud is then used a second time on its journey back to the surface around the outside of the drill string; it acts as a transmission medium to send coded binary pulses back up the hole to an electronic monitor with a digital read-out in the control cabin. 'An elegant solution,' said Blair, momentarily impressed. Then he told me that Ferranti had recently come up with an even more elegant solution: a little electronic package, like an aircraft's black box, that can withstand the battering involved in going down the hole and is accurate, three miles down, to within a couple of feet. 'No doubt about it, to me as an engineer, the most interesting area is the drilling,' said Blair. 'The things they can do miles down at the bottom of a hole are quite incredible.'

Devices like the Ferranti and Teleco tools are needed because the expense of an offshore platform is so great that the wells drilled from it have to cover a very wide area. In the Brent field the reservoir is about nine thousand feet down, but the average length of a well is thirteen to fourteen thousand feet. There were slots for twenty-eight wells on Brent Alpha – I watched them drilling the twenty-third – and they spread out below the platform like the branches of an inverted tree. Alan Jacobs, a former Merchant Navy officer who was then head of Shell Expro's Public Affairs Department in Aberdeen, had explained the problem to me: 'Not only do you want to hit your target in the reservoir, but once you have twenty or thirty wells all starting from the same point, you don't want to drill into another well. The more wells you have, the more complicated it becomes to find your way. If you look at a diagram, you see the wells radiating neatly away from below the platform. But in reality it's like spaghetti. It's sometimes said that the toolpushers offshore are trying to write their names underneath the platforms while they are drilling wells. Can you imagine the torque at the surface when you turn a bit that far away and at such a complicated angle? It would be like laying knitting needles from Aberdeen to London and then trying to turn them. In the old days, of course, there were enormous torque losses, but now we do our deviation drilling with turbines that go down the hole with the bit and are given their correct angle by what we call "bent subs" – "subs" for substitutes – lengths of pipe with angles built into them, like a dog's leg. The Teleco tool enables us to know precisely where and at what angle we are at every point.'

The huge drum on which the heavy steel wire was wound like cotton thread on a reel spun for the last time and the bit came out of the hole. Its top was painted gold, its teeth were silver, clogged with mud and partially chewed up. Compared with the massive machinery all around and the great complex platform whose existence it justified, it seemed a negligible little thing. For a brief moment everyone was still. The bit swung from the end of the wire; the empty hole steamed. Then the chief roughneck spat twice into the hole for luck and rubbed his gloved hands together vigorously, while his companions dragged across the new bit in readiness for the next stage of drilling. Twelve-hour shifts at that pace would be like training for the Olympics.

I followed Blair at the double down to the module deck, a succession of closed areas below the drilling floor where most of the installation's machinery is housed. He heaved back a heavy sliding door that opened on to the wellhead, a low, booming space in which the twenty-two completed wells rose in orderly ranks from the steel floor. On the top of each is a gaudy contraption called a 'Christmas tree', a big tilted cube, chopped from a solid block of steel and painted silver. Red and yellow pipes bristle from the sides. Gauges and dials and taps sprout like flowers from the top and around the base. The coloured pipes from one side of the Christmas tree control the flow of oil or gas from the well. Those on the other side are the 'kill pipes' used to shut down the well in case of trouble; through them is pumped the 'kill fluid', a high-density drilling mud heavy enough to hold down the pressure in the well. When the emergency is over, the kill fluid is replaced with water and in as little as six hours the well can be flowing again.

Each well is lined throughout its length with casings cemented into the strata below the sea. The surface casing, or conductor, at the wellhead is a steel tube, one inch thick and thirty inches in diameter, that looks beefy enough to support the whole platform. These conductors are lowered down through four hundred and fifty feet of water and then hammered three hundred feet into the seabed. The deeper the well goes, the narrower the casings become, starting at thirty inches, finishing at seven. 'A pity it's so calm,' said Blair. 'Because they come straight out of the sea and aren't supported by the platform, they sway around in bad weather. It can look quite impressive. If you don't know what's happening, you think the whole platform is on the move.'

Although one of the engines was out of action, the noise in the turbine module was ear-splitting. Brent Alpha's turbines are driven by gas produced by the platform and each puts out fifteen megawatts of electricity, enough to supply a good-sized village but scarcely sufficient, even combined, to power the installation which is, as Alan Jacobs put it, 'a production well, a refinery, a hotel, and an airport all in one.' Power is needed to run the services on the platform, as well as to drill the wells and drive the machinery that separates the oil from the gas and pushes it along the pipeline to Sullom Voe, the oil terminal in Shetland. It is also needed in even greater quantities for coping with the special problems of offshore production. In the Brent field there are around two thousand cubic feet of gas dissolved under pressure in every barrel of oil – about four times the North Sea average. So for each barrel of oil displaced two barrels of water must be reinjected into the reservoir. Added to that, before the 287-mile gas pipeline to St Fergus was completed, immense power was needed to recompress the gas and drive it back down into the reservoir at six thousand pounds per square inch. According to Jacobs, the four Brent installations alone produce enough electricity to supply Aberdeen and most of the surrounding Grampian region. That day, however, Brent Alpha had to make do with a single turbine that roared like a tortured giant, while two engineers in earmuffs laboured on the other engine and a third workman attacked the steel ladder with a blow torch.

The enclosed parts of the superstructure of offshore installations are made up of prefabricated steel units – called modules – but only those designed for living and working in continuously have proper sound insulation. Elsewhere, noise is a constant condition of life and it doesn't take long to become a connoisseur of its styles and gradations. In the module in which the drilling mud was cleaned the racket was almost as loud as that of the turbines, but broken and dispersed, like an angry crowd on the move. Pumps hissed and banged, engines clattered, giant sieves rattled and shook as they panned the cuttings from the mud. Beneath the steel grille floor a sluggish river of black sludge inched its way into stinking vats. The smell was overpowering, although Blair appeared not to notice it. He stood on a walkway between the shaking pans and the tanks, his feet inches above the sludge, and tried to explain to me the finer points of the process. But the din was too much for both of us, however loudly he shouted. Finally, he turned and trotted off, with me, feeling stupid, at his heels.

In the process module the din was subdued to a steady hum, like chanting. This was the business end of Brent Alpha, the refinery, full of gleaming silver pipes and giant cylinders inside which the gas was separated from the oil. But the only evidence of all that complex activity and the huge pressures was a faint vibration, as though the air inside the module were being compressed from a great height.

The nerve centre of Brent Alpha is a control room like the flight deck of the Battlestar Galactica: banks of dials and lights and digital read-outs, rows of computers, television screens displaying diagrams and figures and abbreviations, all in glorious Technicolor. It seemed hard to reconcile this humming, space-age technology with the agricultural aspects of the installation: the giant hooks and drums, the forest of rust-streaked pipes, the sweating roughnecks and ubiquitous mud. On the dials and gauges and visual display units in the control room the engineers can monitor every function in the production process, from the working temperature of the turbines or the condition of one of their inlet or outlet manifolds to the flow of oil in the pipes and the precise position of the drilling bit two and a half miles down. (Back at Aberdeen is a £30 million master control room to which all the information from all Shell's offshore installations is fed. It looks like a layman's idea of the war room at the Pentagon: charts projected on to illuminated screens on the walls, hushed technicians attending to computer panels and read-outs on which they can monitor every function in the whole of Shell's North Sea fields. If anything goes wrong, engineers on the spot can confer with experts three hundred miles away in Aberdeen with the same information displayed instantaneously in front of them.) But Brent Alpha's cheerful Scots control room supervisor was still not satisfied: 'We've no got telemetry yet. We canna' control the functions from here – only monitor them.' He seemed slightly shamefaced that all his high-tech gear should be, in such a crucial way, limited. 'Still, it takes time to get used to things,' he said. 'When the VDUs were first installed the answers came so quick that none of us believed them. So we always did the calculations ourselves to double-check. Then we realised the computers always got it right. I suppose it would be the same if we had telemetry.'

Outside, there were big gas pipes strung around two sides of the module deck, like a belt of fat. The grid of gas lines from all the installations in the East Shetland Basin was soon to be joined up at Brent Alpha. From it a thirty-six-inch pipeline travels along the seabed to St Fergus, forty miles north of Aberdeen, where British Gas has its receiving terminal at the northern end of the UK gas grid. Littered about the deck were big red plastic diaphragms with rows of steel scrubbing brushes on their concave sides. These were the 'pigs' that clean the pipes, pushed through them from one end to the other by gas pressure. There are 'pig detectors' in the control room to monitor their journey along the pipeline. Blair gave the pigs his supreme accolade: 'They look a bit crude but really they're quite clever.'

I followed him, still at the double, down metal staircases on the outside of the installation until we were about sixty feet above the sea. From up close, the platform's six steel legs, pile-driven into the seabed far below, seemed enormous. Braced between them were giant tubes – painted yellow and scrupulously clean of barnacles and weeds – with railed catwalks strung along their tops. Or rather, along some of their tops: half the catwalks were missing because the sea washes them away as quickly as they are replaced. Above our heads was a maze of girders, crossing and recrossing each other at all angles, like a hedge of yellow thorns eighty yards thick. At their centre, the conductors through which the pipes travelled from the seabed rose straight out of the water and disappeared into the steel floor of the module deck. The sea was deep, deep blue.

In this isolated place – between the water and the purposeful production-well–refinery–hotel–airport above my head – I began to comprehend for the first time the size of the thing and the miracle of its being here. From up above in a helicopter, the installations are dwarfed by the ocean, and once on board you are sucked in by the activity, the detail, the technicalities of the process, and the continual battering noise. But down here, with the intermeshed girders shutting out the sky and the great steel legs dropping hundreds of feet to the seabed and nothing to be heard except the slosh of the sea and the wind whistling, there seemed no limit to the strangeness of the huge structure or the wildness of the place.

In September 1965, the *Sea Gem*, a barge that had been equipped with five pairs of retractable legs and converted into a jack-up rig, struck the first gas in British waters in what was to become BP's West Sole field, off Cleethorpes in Lincolnshire. Three months later, while preparations were being made to move *Sea Gem* to another location, she capsized and sank in rough weather, killing thirteen men. Alex Norquoy, who was born and brought up in Orkney and, like many Orcadians, went to sea as soon as he left school, was at that time a navigator on a Merchant Navy cargo ship. He is a short, tubby man with a red face, black hair, shrewd dark eyes, and a rather gentle manner. His Orcadian accent sounds more Norse than Scots: 'I happened to see the *Sea Gem* when I was heading down to Argentina, and when we came back a few months later I saw her again, upside down with her ten legs in the air. And somehow that sight gripped my imagination. From the marine point of view they were doing things I'd never seen done before: taking what were to me strange vessels, towing them out to sea, setting down their legs, jacking the whole thing up, and then drilling into the seabed; then lifting the legs again after a while and towing them somewhere else and repeating the whole process. All that was fascinating to me as a sailor. I thought to myself, This really looks like something big. I joined Shell a couple of years later when they were just getting under way in Aberdeen and I've worked for them ever since. But I'm still fascinated by it all.'

I had talked to Norquoy in Aberdeen, where he is now movable operations supervisor, one of several managers of a forty-acre site on which Shell stores and maintains the drilling equipment for its sixteen offshore installations. Because I had not yet been out to Brent, the domain he controlled seemed to me fascinating enough. There were warehouses as big as aircraft hangars – painted white and yellow inside and blue outside – housing racks of drilling bits, pressure caps, bent subs, ribbed coupling cones, as well as batteries of electronic gear for testing each piece, since even a minor failure offshore can shut down production on an entire installation. Outside, cranes lumbered around the acres of neatly piled well casings and conductors and carefully labelled bundles of drilling pipes, wired together like asparagus. There were lifeboats waiting to be shipped and countless drums of green and black electric cables, some as thick as my arm, others as thin as a finger. And even outside in the cold wet evening, every exposed threaded end of pipe was greased and wrapped like treasure. This combination of heavy engineering and high precision, of brute strength and delicacy, seemed to me not just fascinating but also, in some curious way, moving, like the scene in the old Frankenstein movie when the monster plays with the child.

Not far off the shore a drilling rig lay at anchor, waiting for somewhere to go. Further out, to the south, a huge installation, lights blazing on its superstructure, was moving slowly out of the rain clouds, a tug straining in front of it. A heavily laden supply boat thrashed out to sea from Aberdeen. A flock of gulls rose suddenly from the edge of the water beyond the warehouses, wheeled in concert, and settled again further up the beach.

That had been my first blurred glimpse of the size and mysterious complexity of offshore oil operations. Now, on a narrow platform below Brent Alpha, with the wind blowing spray in my face, I understood how these gigantic sea-going structures had gripped Alex Norquoy's imagination and never let go.

5

From time to time – usually in emergencies – special pieces of equipment are shipped offshore by helicopter. Everything else goes by supply boat – from the 56,948 tons of piping to the two and a half million eggs, 1,015 gallons of tomato ketchup, and 970 miles of toilet paper consumed each year on Shell's northern installations – a total of more than half a million tons of supplies at a cost varying between £60 and £80 a ton.

Getting there is easy: the journey to Brent from Torry Docks in Aberdeen takes twenty-four hours, from Holmsgarth in Shetland a mere eight. There are only two major problems and they begin when the supply boat arrives at the platform: how to hold the vessel steady in a howling wind and a fifty-foot sea, and how then to transfer the materials to the installation. 'It demands very fine seamanship from the ship's master and very fine crane driving as well,' said Alan Jacobs. 'You get a load on the end of that crane; suddenly it drops fifty feet; the next moment it's coming back up at you. Backloading – transferring materials from the platform to the supply boat – is even trickier. The boat has probably been to other platforms and has containers already stowed on the deck. All that's left is a little square of space and maybe the wind is blowing at forty knots. The crane driver has to put a container down into that little square on the moving deck and let his hook go before

the boat drops again. Then the deck hands have to get it lashed down double-quick because the last three deaths we've had out there have occurred when the stern of the boat has tipped up and an unsecured container has gone wallop and squashed the poor bastards flat. Maybe half a million tons of supplies doesn't sound much, given the size of our North Sea operations. But when you consider the difficulties simply of moving the stuff on and off the platforms you get a different perspective.' Like nearly everyone involved in the offshore world, Jacobs enjoys dazzling outsiders with the telephone-number statistics of the business and the disproportionate troubles that have to be overcome. But in some areas he, too, remains impressed. 'Those crane drivers,' he said. 'They're real artists.'

The glass-sided control cabin of the crane on Brent Alpha is close to the flame of the gas vent that sticks out horizontally over the sea and warms the crane driver, whatever the weather. 'It's nice and cosy now,' said Gordon Bunch. 'But in summer it gets a bit torrid.' Gordon Bunch is a corpulent, middle-aged man with curly hair, a soft West Country accent, and twenty years' experience as a crane driver – five of them in the North Sea. His generous belly comfortably filled the space between the cabin's padded chair and the row of little levers that controlled the crane. He played them like an organist while I stood outside on a railed catwalk with a hundred and fifty feet of space between the steel grille of its floor and the churning sea. 'Your elements,' he said when I asked him the difference between working offshore and on land. He swung the crane around until our backs were towards the gas vent and the hook was above a little slot on the platform's superstructure. 'Wind makes a hell of a difference. We reckon to work up to forty knots. Occasionally, we have to go over, but it's Shell policy to stop at forty. And more than that, it's not ideal, like. You haven't got the same control. The containers spin like tops when you pick them up.' A foreman stood beside the slot and patted his hands up and down as if he were playing bongo drums. Bunch moved a lever and the hook sank out of sight into the slot. He went on talking: 'Besides that, the boat is moving up and down, see, she's moving from side to side. Plus the fact that if they can't hold her off, she's going in and out. And that's where your precision comes in. You've got to hold your load over, then drop her down. But providing it's not too rough, you can square it up pretty well to where they want it.' The foreman played another brief riff on his invisible drums, Bunch juggled his levers, and the hook came up with a rubber flotation bag hanging from it. Beneath the bag dangled curved pieces of steel tubing that stuck out in all directions. The load rotated in the wind yet cleared the slot with inches to spare and touched nothing, despite the fact that Bunch was still driving blind. He nodded approvingly at the foreman and said, 'He's a good man for directions.' He swung the load through two hundred degrees and dropped it gently on to a deck, just out of sight, where the platform's spare equipment was stored. 'It's a matter of coordination with your levers, really. Just your basic crane skills which you've learnt over the years – only more so. In actual fact, sometimes it's quite fun when the wind's tricky and the boat's playing up. But the real difference out here is height. It doesn't matter how long you've been driving cranes, when you first come offshore and drop that hook down you're likely to be miles out. But once you get used to it, you drop your boom out and you know more or less exactly where the hook is going.'

A supply boat had drawn up to the platform and was pitching up and down in the boisterous sea, waiting to unload. It was a clumsy-looking vessel, its bridge and crew's quarters squeezed on top of each other in the bow, leaving a great flat platypus tail, crammed with containers. It was brand new. 'Not a week old,' said Bunch authoritatively. Its paintwork was gleaming terra cotta; its wooden deck was unmarked. *'Baldar Vigra, Oslo'* was painted on its stern.

'I'll show you what I mean,' said Bunch. 'I'll swing out the boom and you tell me when to drop her.'

I leaned into his cabin and peered through the window while the crane rotated. The *Baldar Vigra* looked improbably small and far off but I thought it was worth a try. 'Now,' I said.

Bunch pulled a lever and the hook dropped. It landed sixty feet short of the boat's stern. 'See what I mean?' said Bunch.

He brought the hook back up, manipulated the levers again, then lowered the hook gently until it was directly over a container in the stern. Two sailors, as agile and quick as monkeys in the confined space, attached the hook to cables on the container. They waved. Bunch waited, playing his line like a fisherman, watching the waves and the rolling, bucking boat. It rose and fell, rose and fell, rose again. At the precise moment when the boat was at the peak of the incoming wave, Bunch pulled a lever and the container slid clear of its narrow space. It swung up and around in a great arc, like a cannonball, then stopped suddenly, as if by word of command, a foot or so from Brent Alpha's superstructure.

'Magic,' I said.

'Ah well, see, that's where your crane driving comes in,' Bunch said. 'The trick is to keep up with her, so you're always over her, never pulling, never letting her get ahead. It's what you call a knack.'

The foreman had moved round and was now playing his drums again above the invisible equipment deck. Bunch watched him, not his levers, and the container sank out of sight. A few minutes later the foreman signalled again, Bunch manipulated the levers, and another container rose into the air from the equipment deck, swung smoothly up and around, stopped suddenly in midflight, and dropped into the first container's slot on the heaving deck of the supply boat. The deck hands uncoupled the hook and waved. The moment the hook was clear, the *Baldar Vigra* took off for the next stop on its schedule.

'Jesus,' I said.

Bunch looked pleased.

6

The next morning the sky was overcast, a cold wind was blowing, and all the early shuttles were full. But there was room for me on the eleven o'clock bus that was taking a doctor and medical orderly out to Stadrill. The doctor, who was new to the job, was touring the installations to see what their sick bays had to offer by way of medical facilities. The orderly, an old hand, was going along as his minder, ostensibly to check that all the first aid equipment was in place. The Shell field's hospital is on *Treasure Finder* and is fully equipped with curtained beds, an operating table, and all the necessary, ominous gear. Elsewhere the facilities are more basic and anyone needing serious medical help is flown either to *Treasure Finder* or back to the beach. The doctor was a tall, blank-faced young man from the Transvaal, just out of medical school, who spoke – briefly – only when spoken to.

Stadrill, that month, was at the northeast corner of the field, beyond North Cormorant and very remote. The eleven o'clock bus took its time, calling at most of the platforms on the way with passengers and freight, dropping off and collecting, doubling back from Bravo to Alpha with a single passenger and a small parcel, then off at last over the choppy sea northwards, with no landfall before the frozen archipelago of Spitzbergen, up on the seventy-eighth parallel, and the Arctic ice cap. I leaned forwards, hoping to see another installation or at least a supply boat. Nothing. I remembered what I had been told about the feeling of apprehension that accompanies every helicopter flight. I also remembered a couple of uneasy lines by John Berryman:

> Starts again always in Henry's ears
> the little cough somewhere, an odour, a chime.

The South African doctor stared out of the window with a face like stone. The last leg of the flight took twenty minutes but seemed far longer. With no gas flare to announce its presence, Stadrill looked small and lonely in the endless grey water.

Stadrill is an exploration drilling rig, with no processing plant and a crew of a mere seventy to eighty, half the complement of even a small installation such as Brent Alpha. Like *Treasure Finder*, she is a semisubmersible – a coffee table on two submarines – and she moves under her own power from one drilling site to the next. She is old, as North Sea vessels go – built in 1975 as a replacement for Staflo, the first drilling rig in the area – but the accommodation is comfortable and the cabins, even those with four berths, are considerably more spacious than those on *Treasure Finder*. Because she is the one installation in the North Sea that is crewed and operated wholly by Shell, with none of the work contracted to outside firms, those on board are the longest-serving Shell employees in the area; some had graduated from Staflo and had been shuttling back and forth to the beach for as many as fifteen years. The atmosphere was friendly and relaxed, with none of *Treasure Finder*'s bustling, transit-camp impersonality.

Stadrill has two OIMs: the barge engineer when she is on the move, the toolpusher when she is stationary and at work. The toolpusher, Peter Carson, was bearded and lanky, with narrow-set eyes, a neo-Australian accent – he had worked in Brunei – and a limitless fund of expertise. He rattled off the fine points of directional drilling, the chemistry of drilling mud, the intricacies of core sampling, and the differences between FINDS (the Ferranti Inertial Directional Survey) and the Teleco tool, leaving a blur of high tech and a strong, precise impression of a lover's passion for his subject. But when we toured the rig the detail was blurred even more by the terrible smell. The mud they were using on that particular section of the well was based on diesel oil and the whole platform reeked of it in varying degrees: bad on the drilling floor, intolerable in the module in which the mud was cleaned and recycled. Outside the living area, the mud coated everything: the decks, the stairs, the handrails, the machinery; within minutes it had coated us, too. A black, padded executive chair – plonked down incongruously in front of the Teleco tool read-out panel – was encrusted with the stuff, as though someone had rolled in it and then sat down heavily.

Standing on one side of the high deck below the drilling floor was a gigantic blowout preventer, a device that controls the flow from the well, like the Christmas tree but many times larger. It would be used eventually to cap the wildcat exploration well the crew were working on above, but for the time being it was dismantled for repairs. Two technicians were squatting inside it, as though in an elongated diving bell. One of them, wielding an outsize wrench, wore thick, scholarly glasses and what looked like a stethoscope around his neck.

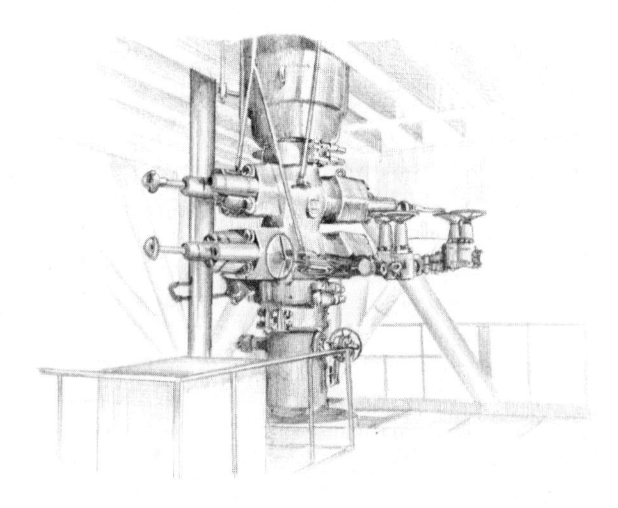

Down here the smell of diesel was almost bearable because at the centre of the deck was the moonpool, a great square space opening to the sea. And at the centre of the moonpool the riser from the well came out of the heaving water and disappeared into the derrick floor above. Steel cables dangled around it, complicated girders and struts protruded at all angles, and the whole thing seemed to be swaying dangerously, although the riser in fact was stationary and it was Stadrill that was shifting on its invisible pontoons. When I leaned over the guardrail at the moonpool's edge to breathe the fresh air I could see the waves building up threateningly around the legs of the rig.

Stadrill's barge engineer was a wiry little man with a pointed Beelzebub beard, sharp dark eyes, and a taste for irony. His name was Martin Law and he had spent thirteen years in the Merchant Navy. 'I started in passenger ships and went downhill from there,' he said. 'Ended up in a tanker.' His control room was crammed with shining banks of electronic gear, but since it was also the ship's bridge there were windows along one side with a view of the empty sea. At their centre, facing a great bank of read-outs, switches, and levers, was a comfortable raked chair in which he sits when Stadrill is on the move. He told me he had first come out to the North Sea oil fields in the early wildcatting days when safety regulations came a distant second to the action. He had been mate on a supply boat carrying a container that was urgently needed by an American exploration rig. A storm was blowing, the seas were high, and the winds were gusting far above the forty-knot limit now imposed by Shell. But the Americans had wanted the container very badly and the crane driver and deck hands knew their jobs. Somehow or other they managed to hook the crane to the shackles of the four cables attached to the corners of the container. But just as the crane driver was about to lift the thing off, two vast waves hit the supply boat and drove it off. 'The crane looked like a fishing rod that's been taken by a shark,' said Law. 'The cable was almost horizontal.' While the captain fought to bring the supply boat back close to the rig, the deck hands tried to release the hook. 'But the cable was as tight as a fiddle string. It was too dangerous. So the crane driver did the only thing he could: picked his moment and yanked the container up. It went flying into the air like a football. Two of the shackles broke but the other two held, thank God, and no one was hurt. I thought, Enough is enough. It's time I joined the other side.'

The control room was calm and hushed; the only sounds were a faint hum and an occasional click. But outside the window – distanced, as in a silent movie – the wind had got up and the sea was steadily rising. A small trawler ploughed slowly into view, crossing from right to left. Every installation, by law, has to be attended by a standby boat that circles constantly, like a sheepdog around a docile flock, in case a man falls overboard or the platform has to be abandoned. The standby boats work four-week shifts – a monotonous job but at least a steady source of alternative income for the economically depressed Scottish fishing fleet. The standby boat outside the window moved so slowly that she seemed to progress hardly at all. But she rolled recklessly from side to side and at the top of each wave her bow lifted clear out of the water. 'That's nothing,' Law said dourly. 'In a big sea she'd go down so far you wouldn't see her mast.' A couple of months earlier, he told me, the standby boat had been out in front of Stadrill when a violent storm blew up. She was headed into a wind so fierce that she was unable to turn and had steamed sixty miles north before the storm blew itself out, although, according to the rule book, she is never supposed to stand more than five miles off. Law had watched the pantomime from his captain's chair, with Stadrill rolling no more than a few degrees at the height of the storm, and he told the story now with relish. 'Look at the bugger,' he said contentedly. 'You'd need a stomach of corrugated iron to do that job.'

Because of the deteriorating weather, the three o'clock bus was cancelled. The doctor, the orderly, and I sat in the recreation room, watching *Assault on Precinct 13* and drinking instant coffee out of plastic cups, until the five o'clock shuttle arrived. The sea was dirty grey and bristled with white horses and the clouds raced low, blotting out the horizon. The chopper rattled along, in and out of the bottom layer of cloud that sank lower and merged into mist as we went south. By the time we reached *Treasure Finder* it was almost dark. Brent Bravo and Charlie were blurred shapes hung with lights, Alpha was a shadow, and the remote gas flare a wavering banner of fire, disproportionately large, signalling wildly in the darkness. When the helicopter touched down softly on *Treasure Finder* I felt as if I had come home.

That night the field was so full that eleven men had to be flown back to the beach because there were no beds for them.

7

Before I went offshore I had been told that, whatever the other drawbacks of life in the North Sea, at least the food was what the *Guides Michelin* would classify as 'worth the detour'. 'French chefs,' they had said in Aberdeen. 'Haute cuisine. Don't be fooled by the fact that the lads get through a thousand bottles of ketchup a year. They can't help themselves. It's a conditioned reflex.' On the spot, the lads themselves were less impressed. 'If a man's dissatisfied with any aspect of his life out here, the first thing he complains about is the catering service,' said Ken Murray, Brent Alpha's OIS. 'Obviously, we can lay it on when we like and as a general rule the food is fairly good. But I personally don't feel it's quite what it's cracked up to be. The stories you hear back on the beach about the marvellous nosh we get are simply not true. Certainly, there's a menu for every meal and it's always substantial, but it's not like a night out on the town with the wife.'

More like an evening at home with the wife, perhaps. And that is appropriate after a killing day's work: mammoth helpings of roast meat and two veg; big, overcooked steaks on demand. It is the spirit of the traditional Sunday lunch, the best possible prelude to sleep. The breakfasts, however, are sensational. The cold table in *Treasure Finder*'s mess hall is piled with fruit and jugs of juice and as many brands of cereal as a well-stocked supermarket. (Later in the day it is spread with salads, pickled herrings, and a vast assortment of cold cuts.) As for the hot food: the cheerful kitchen staff ladle out porridge, bacon (four or five rashers at a time), sausages, black pudding, kippers, haddock, and as many eggs as you can handle (fried, scrambled, poached, or boiled). There is coffee (real, instant, decaffeinated), tea, hot chocolate, milk and great jugs of cream, bread (white and wholewheat), rolls, and toast, and a small regiment of sauces as well as the ubiquitous ketchup. The variety, like the quantity, is overwhelming.

It needs to be. Without alcohol or live television or current newspapers – the latest are rarely less than a couple of days old – food is the only form of relaxation, apart from videos, and the mess hall the only place where people pause to talk a little before weariness and digestion engulf them. It is also the place for what passes offshore for jokes. One evening over dinner, an enthusiastic, hard-sell voice announced over the PA system, 'Just a few tickets left for the Saturday night disco on Statfjord. Apply now to Admin. First come, first served.' The man sharing my table was built like a gorilla. He wore a black sleeveless T-shirt and both his arms were tattooed from shoulder to wrist. He winked, shook his head, sighed. 'Take nae notice,' he said. 'There's a mug on board fresh out from the beach who doesna' know his arse from his elbow. The lads are trying it on. Same old game. Always Statfjord. Foreign, you see. Different rules. They tell him there'll be beer and big blonde Norwegian stewardesses to dance with. Last month some puir wee chappie swallowed the whole thing hook, line, and sinker. Dressed himself up in his Sunday suit, went up to Admin – Arthur up there goes along with it – put on a survival suit, and waited two hours for the chopper before he saw the joke.'

The morning after I went out to Stadrill, the wind was still blowing and the sea was high, yet the whole field was blanketed in fog. Because the area is so vast the forty-knot fog is a quirk of the North Sea oil fields, swaddling and blowing at the same time, grounding the helicopters and chilling you to the bone for hour after interminable hour. In summer, when warmer air clashes with a cold east wind, the haar comes down, a fog like smoke that immobilises everything for days on end and drives the men who are waiting to fly home crazy with impatience. 'There's a ten-mile fog bank approaching,' Arthur the Admin clerk announced one afternoon to the roughnecks who had checked in for the last Chinook back to Aberdeen. 'And that's just its height.' Nobody even smiled.

But a gale-force fog is, by nature, a short-term threat and the atmosphere that morning in the mess hall was relaxed. I had arrived for breakfast at seven – late for *Treasure Finder*, where the meal is served from five a.m. until seven-thirty – and lingered over coffee, chatting to a young scaffolder from

Aberdeen whose name was Roy. The scaffolders have one of the most spectacular of the North Sea jobs, almost as dangerous as that of the divers but more visible. They work mostly on the network of girders beneath the module deck, repainting them constantly in bright Shell yellow as a protection against the weather. The girders overhang the churning sea, and the main working and living areas of the installation overhang the girders, isolating them in a world of their own. In that lonely space, the scaffolders erect their steel spiders' webs, then climb out over the cold, empty space, sandblast the old paint away, prime the surface, and paint it over twice. Even for those with a head for heights, the loneliness of the situation and the brutality of the weather make North Sea scaffolding as truly a 'dreadful trade' as that of the samphire gatherer in *King Lear*. 'You've got to admire the scaffolders,' said one of the roughnecks. 'But I don't much like them. They're all prima donnas.'

Roy from Aberdeen, however, was modest and rather shy. He had a narrow, bony face and when he talked his head came forwards and the skin stretched tight over his cheeks and neck as though he were straining at a leash. He said it had been a year since he was last offshore, far too long to wait, he couldn't get enough of it.

'Because of the pay?' I asked.

'Oh, the pay.' He ducked his head. 'At night you look around at the flares and lights and you see all that money.' His voice was low and full of awe, like that of a man in church. 'Of course the pay's good, but that's not the point. What I like is the peace and quiet, the fresh air. You get away from the girlfriend and the family, from all the troubles. It's like going through one of those black holes from one universe to another. A few months ago, I did a job on the west coast of the Highlands, north of Ullapool. It was the same there: peace and quiet and fresh air.'

A large, grizzled man, sitting at the end of the table, raised his head from an Alistair MacLean thriller and stared balefully at the young scaffolder. 'Fresh air!' he said. 'You go out there on deck and nine times out of ten you get a face full of soot or acid-grease fumes. Three quarters of the people on board go home with dandruff and God knows what other skin complaints. Fresh air, he says!' He went back to his book. The foghorn sounded its three long notes.

The scaffolder blushed. 'Better than Aberdeen, for a' that,' he muttered. Then he launched into a story about an accommodation barge that had lost its anchors in a storm during the winter of 1982 and drifted for three days with all six hundred men on board singing 'Anchors Aweigh'. 'When the rescue boat finally got a line on to her the first man off was the safety officer!' Like Martin Law's story about the safety boat that could not turn, it was typical of North Sea folk tales, all of which serve one purpose: to sanitise disaster by turning it into a joke. That, too, is part of an elaborate but unspoken code of behaviour that everyone follows in order to make tolerable the constricted, precarious life offshore: never dramatise, never dwell on the possible dangers, never intrude.

One evening I took my plate of food to a table where two Geordies were sitting – heavyset men with grey, drained faces. One of them had his right hand swaddled in bandages. He said he had slipped on a greasy ladder, fallen, and cut his thumb to the bone just above the tendon that joins it to the palm. His eyes behind his thick spectacles were dim and bewildered. He ate left-handed and with difficulty. His companion asked, 'Does tha food want cutting up?' The injured man shook his head. 'Nay, I'll manage.' He shovelled roast beef awkwardly into his mouth, then said more cheerfully, 'I gotter see t' quack again tomorrer. Maybe he'll send me back to t' beach.' The other man nodded sympathetically, watching him carefully as he fumbled with his food. He seemed as worried and protective as a mother with a sick child.

Later, when I was on Brent Charlie, I asked the OIM, a soft-spoken former Merchant Navy officer called Bob Ingram, if he thought life offshore might not be easier if there were women – nurses, for instance – working alongside the men. He shook his head vehemently. 'You get some really tough cases out here, men you wouldn't want to get within a mile of if you saw them in Aberdeen with a couple of drinks inside them. But offshore, if they get in each other's way, they say, "Sorry, mate." Women aren't like that. Women aren't polite.' He waved a hand vaguely, as if to forestall an argument. 'Don't misunderstand me. I'm a married man – happily married. But conditions out here are difficult and the work comes first. You have to learn not to take things personally. You have to keep your distance. In my experience, women don't find that easy.'

Work and impersonality are the rules of the game up there. The complexities of oil exploration and production are infinite and their solutions are imaginative, subtle, even delicate. But life itself is simplified, absolved from emotional ambiguities. Yet it is also curiously – to use Bob Ingram's word – 'polite', with none of the abrasive us-versus-them, workers-versus-management hostility that plagues British industry back on the beach. Perhaps that is because there is already quite enough hostility out there in the stormy North Sea. Or perhaps it is because most of the workers are specialists of one kind or another, whose emotional lives are absorbed by their work. Or perhaps, after a twelve-hour shift, they are simply too tired for anything except peace and quiet. Whatever the reason, the atmosphere in the mess hall and coffee shop and recreation rooms is subdued, friendly, undemanding.

The helicopter hangar on *Treasure Finder* is brightly lit, as clean as an operating theatre and permeated by the clear, sharp smell of engine oil. There were no flights on the morning of the forty-knot fog and the four Bristow Bell 212s were lined up side by side, their rotor blades tied back along their fuselages. In the shadowless, fluorescent light their red, white and blue liveries and vertical rear rotors, painted white with a narrow red stripe, looked as bold and cheerful as a pop art painting.

On one side of the hangar two mechanics were working on an engine. One was a Yorkshireman with fair hair and sharp features, the other a small, dark man from Yarmouth who had learned his trade in the RAF, then joined Bristow at the time when gas was discovered off the coast of his native Norfolk. They worked like surgeons, unhurriedly, with great precision and concentration, wiping clean each part of the engine as they dismantled it and placing it neatly on a shelf.

The Yorkshireman said, 'You only get helicopter work in the armpits of the world. For ordinary conditions, choppers are too expensive.'

'God, when He worked it all out, said, "If they want oil, they're going to have to look for it," ' added the man from Yarmouth.

They said they had worked in Singapore, Iran, the Persian Gulf. 'I like a change,' said Norfolk. 'A couple of years here, a couple of years there. It's a bit of a gypsy's life. So okay. But I couldn't work a nine-to-five; it'd drive me mad.'

They told me they did two-week stints offshore: night shift the first week, day shift the second. Norfolk said, 'In winters the days are so short up here that when you work night shifts you don't see daylight for a whole week. It's dark when you come on at six and dark when you finish twelve hours later.'

Yorkshire was holding a screwdriver between his teeth while he used both hands to fiddle with something small inside the engine. He removed the screwdriver. 'Could be worse,' he said. 'Some of the roustabouts work the night shift for two straight weeks. When they get back home their bodies are on Australian time and it takes them days to adjust. That's so much real time lost.'

I said, Surely their experience and expertise qualified them to work where they wanted, not just in the uniformly dreadful places where oil is drilled. The Yorkshireman answered, 'A helicopter's an expensive piece of kit, compared with a fixed-wing aircraft. The operating costs are very high because there are so many moving parts – rotors, gearboxes, things that can go wrong and have to be replaced. So a company like Bristow's can't rely on casual private passengers; there aren't enough VIPs to go around. Only the oil companies have that kind of money and need that kind of regular work.'

'Anyway, who wants VIPs?' said the man from Yarmouth. 'Out here something's happening, they're getting results. You can see it, you can hear it, and most of the time you can bloody smell it. There's a satisfaction in that, you know.'

'Too bloody true,' said the Yorkshireman.

The fog blew all that day, so after lunch I crossed the caged-in gangway – known cheerfully as 'the widow-maker' – that joined *Treasure Finder* to Brent Delta, the three gigantic concrete legs surmounted by eight levels of decks and living quarters that towered over the flotel like a monster from outer space. Brent Delta is a concrete gravity platform, 228,231 tons in weight and 981 feet high – three feet shorter than the Eiffel Tower – from its base to the tip of its vertical gas flare. It rests on the seabed in 466 feet of water and resists any storm by virtue of its immense weight. Around its legs, far below the surface, are ranks of cylindrical storage tanks with domes as large and smooth as those of an Islamic mosque.

Andy Wood, the Shell toolpusher on Brent Delta, is a slight, wiry Lancastrian with a blond beard and intense blue eyes that make him look like D. H. Lawrence. The walls of his office were plastered with charts and plans of the forty-eight wells that will eventually radiate from the platform. Filling the spaces between them was the usual assortment of nude photographs cut from calendars sent out by oil equipment companies. These photographs are in every offshore office apart from those of the OIMs – who are presumably too senior for frivolity – and Admin, where every available inch of wall is crammed with lists of names for accommodation and transport. But the smiling faces and glowing bodies seem to say less about desire than about the loneliness of life out here in the North Sea. Unlike those elsewhere in the field, the nudes on the walls of Andy Wood's office were framed – as a token, perhaps, of his seriousness.

On the way down to the drilling floor he grumbled routinely about the vertical gas flare, blazing far above our heads: 'When the wind's southerly it blows the flame over the top of the living quarters,' he said. 'The ventilators suck it in and everybody bakes. Then a wind from the north goes straight into the ventilators and we all freeze. And there's nowt they can do about it.' But he spoke without conviction, more as something to say than as a genuine complaint.

Down on the derrick floor, where the drill roared and spun in the rotary table, he began to lecture me on drilling mud and his manner changed: he became precise, technical, yet curiously passionate, as if this were one of the things he cared most about in the world. He talked about the subtle chemistry of the mud and how critical its weight becomes as the bit moves through different formations; the deeper it goes, the heavier the mud must be. 'Look,' he said. 'The hole we're on now is two and a half miles deep and the pressures down there are enormous. Even so, it takes the mud two hours to travel back from the bit to the surface. Two hours, reet. Now do you get the idea?'

'I suppose so,' I said, although all I could see was the intricacy and huge scale of the operation: not just out here in the North Sea, and not just the extraordinary technology that has been evolved for the job, but the job itself. 'The earth has cut an artery,' said Dusty Woods when the Lake View well in the Coalinga field blew out in 1910. He was right. Oil exploration, like the exploration of space, deals in a larger dimension of reality than most of us know how to cope with. Beyond the technical sophistication and sheer hard labour is something far stranger, something that has to do with the anatomy of the planet itself. In *The Wildcatters*, Samuel W. Tait listed some of the wilder theories of the origins of oil:

> One philosopher argued that the earth was a huge animal, the water its blood, the rocks its bones, the grass and trees its hair, the hills pimples on its face, and Aetna and Vesuvius merely eruptive boils; oil could naturally be secured by boring through the skin into the blubber of the immense creature. Another . . . [declared] that petroleum was the urine of whales conveyed from the North Pole in underground channels. A Methodist revivalist conducting a camp meeting in the 1870s said that petroleum was stored away to be used to destroy the world on the inevitable dreadful day. Perhaps this was the same one who asked the legislators of an oil state about that time to prohibit further drilling because oil was used to light the fires of Hell, which might

go out if petroleum continued to be brought to the surface for a profane purpose.

Out here on Brent Delta, with an experienced technician patiently describing to me the logic of the process, all I could think was that the dotty theories that had once tried to explain the mystery of oil were, after all, emotionally justified.

The rock Andy Wood's crew was drilling through was hard and progress was slow, so no pipes would be needed for the next hour or so. The little board high up in the derrick which I had first noticed on Brent Alpha was unmanned. I pointed up to it. 'What's that called?'

'The monkey board.'

'And the guy up there, what does he do?'

'The derrick man.' Wood looked me up and down critically. His blue eyes glinted. Briefly I saw myself as he must have seen me: another time-waster from the beach, grey-faced, grey-haired, with a notebook in one hand, a pen in the other, and a tape recorder sticking out of the pocket of his brand-new overalls. 'Got a head for heights, have you?' he asked. 'Want to go up?'

I looked up at the greasy cables and the great hook moving in the wind. 'Why not? Part way, at least.'

On the outside of the derrick was a vertical steel ladder, its rungs plastered with oil and mud, slippery as ice. We climbed up to a platform fifty feet above the drilling floor, halfway to the monkey board. A foggy wind howled through the latticework of the girders. We were already above the helideck that sits on top of the living quarters, eight decks plus a hundred or so feet of space above the sea. Painted on the helideck was a big yellow circle with a yellow line sticking out from each side of it like an axis and a white 'H' at its centre. On one side of the deck was painted 'Brent D 211–29'. I thought, This is how the helicopter pilot sees it as he comes in to land; it's not very big. Off to the right, *Treasure Finder* lay far below us, diminished and foreshortened by the distance. It seemed to tilt heavily from side to side in the swell, although on board there is only the vaguest sensation of movement and the one clear evidence that the flotel is not fixed to the seabed is when an unlatched locker door in the cabin swings slowly open and shut.

Wood and I walked around the platform, peering out into the fog. Only Brent Charlie was visible, a blur of lights surmounted by a ponytail of flame. The foghorn boomed and a small hard rain came down, soaking through my overalls and bouncing off my plastic helmet. We stuck our heads through the girders and craned upwards while Wood explained the derrick man's work. Then he broke off abruptly and said, 'Might as well see for yourself. Let's go on up.' I followed him carefully up the next forty feet. My boots skated on the rungs; the mud oozed through my gloves.

Up at the next level there was a narrow platform; around the inside of the derrick metal sheeting helped to protect the monkey board from the weather. The board itself jutted out about four feet from the side of the derrick, narrow as a diving board and soaked with oil. At either side and at right angles to it, fingers of heavy steel, six inches apart, protruded from the sides of the derrick, ending a few inches short of the board. Along the length of the fingers were short steel bars, hinged so they could swing through one hundred and eighty degrees from one finger to the next. A few stands of drilling pipe leaned like rusty saplings, their tops protruding from between the fingers, each separated from the next by the little steel bars. Dangling from a rail behind the monkey board was the derrick man's safety harness: a leather belt attached to a length of hemp rope. He leans out on it from his greasy little platform, grabs the top of one of the ninety-foot lengths of pipe racked between the fingers, slides it along the slot next to the monkey board, stabilising it with a rope, and throws it into moving 'elevators' – big clamps with handles – that latch automatically when the pipe is in place. The elevators lock on to one end of the pipe, the roughnecks far below take hold of the other end, and together they heave it across to the rotary table and screw it on to the drill string. On a round trip, when the bit is being pulled back up the hole, the process is reversed. The derrick man lassoes the end of the pipe and the clamps are released.

He then pulls it along the gap beside the monkey board, slots it into place in the fingers, and flicks over the little steel bar, which locks the pipe into position and, at the same time, opens up a space between the adjacent fingers. A simple, logical procedure, given sufficient muscle.

'Like tossing the caber,' I said.

'Sort of,' said Wood. 'Only the pipe bends, see. So the trick is always to go with the swing of the curve. If you pull against it, it's too heavy to move.'

Up here, as everywhere else, all the surfaces were greasy and wet, but Wood moved around nonchalantly, as if the stubby monkey board and slippery fingers were as wide as a lawn, and as if there were no gaps between them. He seemed to be enjoying himself. On the drilling floor below, the roughnecks were reduced to yellow hats and oily arms and shoulders. The rain pattered against the metal sheeting; the wind whistled in the girders above.

We went out on to a railed catwalk on the outside of the derrick. At one corner was a shallow wooden box containing a parachute harness, a coiled rope, and a friction clutch. 'For emergencies,' said Wood. 'If the derrick man can't climb down, he can strap himself into that and jump.' Near it was a smaller box in which hung a breathing mask and a cylinder of oxygen. I asked if they had ever been used.

'Not so far. But we check 'em once a week just in case.'

We walked around the catwalk and admired the fog. A supply boat, loaded with pipes, was heaving up and down far below, seemingly attached to the installation by the crane, like an aircraft refuelling in flight. From four hundred feet up it looked tiny. Brent Delta's decks were a jumble of machinery, containers, and stacked pipes, with pin men in yellow helmets moving busily among them. To our left, the steel trellis of the gas flare rose above us with a chivalric pennon of flame flapping into the fog.

'What a place!' I said.

'You get used to it.'

'Speak for yourself.'

We went back down the slippery ladder with the wind

tugging at us and the rain stinging our faces. When we reached the derrick floor Wood grinned and said, 'All right.'

'My pleasure.'

'You want to go down the inside of the leg?' asked the production engineer.

I was in Brent Delta's recreation room, stretched out on a sofa, boots off, feet up, sipping a cup of coffee. The rain had soaked through my overalls and North Cape jacket; there was a layer of damp between my shirt and my back. Well, I thought, at least it will be out of the weather.

The production engineer was a tall man with a quiet voice and a peaceful manner. He had a ruddy complexion and very regular features. His name was Terry Smart.

'I thought they were full of oil,' I said.

'The oil's in the tanks down below. Two of the legs have well pipes in them; the other houses pumps and things.' He eyed my soaked, stained clothes sympathetically and said, 'Don't worry. There's a lift.'

I followed him down through decks full of machinery to a platform at the top of the leg. Beside the lift doors was a white plastic board on which we wrote our names with a coloured marking pen. Smart handed me a pack like the one I had seen on the derrick, containing breathing equipment. 'Just in case something goes wrong with the ventilation,' he said. 'Not that it ever does.'

While we waited for the lift, I went over to the railing on the opposite side of the platform and peered through a rope mesh down the great central shaft of the leg. It was like laying my ear to the cone of a loudspeaker when the volume is turned up full: a great deafening column of sound, sporadically lit, dwindling into darkness. Far below was another rope mesh; below that, glinting, inky water. Once again, I felt dwarfed by the sheer size and complexity of the project – of which this, after all, was just one small part.

The lift doors opened and Smart pressed the bottom button. My ears popped regularly as we dropped.

We came out at the seventy-one-metre level, exactly half

way between the bed of the sea and the surface, into a big circular chamber, seventeen metres in diameter. The walls of the leg were grey concrete, one metre thick. I lay my hand flat against the dank surface. Beyond it was the North Sea, black and utterly still at this depth, stretching up to the Arctic. All that darkness and silence were hard to imagine in the din of the machinery and it seemed odd to be moving freely when a few feet away the pressure would weigh like lead.

The chamber was crammed with roaring booster pumps that drive the crude oil from where it is stored in the domed tanks clustered around the installation's legs and up a huge pipe that eventually leads it from Brent Delta to Sullom Voe. Next to this pipe were three smaller pipes in which the oil – separated from its gas in the process plant on the module deck – was fed into storage. The lift shaft dropped free from the upper deck to the seventy-one-metre level, a narrow steel staircase spiralling around it, making it look like a gigantic corkscrew. Opposite the lift doors rose a great square ventilation trunk and all around the walls were narrower pipes and dense lianas of electric cable, each encased in heavy black rubber. These cables are everywhere on the installations, outside as well as in, climbing vertically and disappearing mysteriously into holes, or running horizontally in shallow troughs around all the modules on deck, like some post-modernist architectural motif. 'Beats me where they all go,' said Smart. 'But they're all doing something.'

Beneath the steel grille of the deck was a lower deck, housing yet more drumming machinery. Below that was the dark, oily water of the bilges. When I looked back up the square shaft at the centre of the chamber I could see three more decks of pumps, diminishing lines of pipes and electric cable, an irregular pattern of spotlights and fluorescent strips. It was like looking upwards from the bottom of one of Piranesi's imaginary prisons – a vast, enclosed, shadowy place, with gangways and galleries and ominous, purposeful machinery, all of it disproportionate to the human scale. A prison without people, however. The machines went about their noisy business by themselves, untended, monitored by other machines with men sitting at their consoles up there in the busy world. Drops of water were falling from somewhere far above. One of them hit me square on the forehead like a pellet from an air gun. I pulled my head in sharply.

'We'd best be getting back,' said Smart.

As the lift rocketed up towards the top of the leg, it seemed to me there was no end to the strangeness of life out here. In the space of one hour I had been four hundred feet above the rowdy surface of the sea and two hundred and twenty-five feet below it. In order to understand the offshore world you have to revise your ideas of geography and distance. The installations are like mountains in a remote Alaskan range: you reach them by air; after that, the significant distances you travel are vertical. But even stranger was the way both my guides took all this for granted. To be poised, working, far above the waves or moving about among engines deep below the surface was, for them, merely part of their daily routine. But perhaps the only way to function in an environment as hostile and outlandish as that of the North Sea oil fields is to absorb yourself in the details of your work and cultivate towards everything else – the place, the weather, the scale of the projects, and the continuing improbability of the situation – a busy, sane indifference.

8

There is a colony of about six thousand Americans in Aberdeen, representatives of Texaco, Amoco, Conoco, Occidental, and the numerous specialist companies that provide services, staff, and equipment to the oil companies. The Americans have their own school and their own clubs and – by Scottish standards – they tend to overheat their houses, but mostly they have settled easily into the austere, wealthy granite city. The young ones patronise plush bars and discos like Mr G's and Champers; the more sedate join the Petroleum Club, which offers a swimming pool, squash courts, golf and evenings of country-and-western music; the most senior, depending on the season, fish for salmon or murder grouse with the local gentry. One or two have even bought and renovated their own Highland castles.

Yet although they may temporarily give up bourbon and develop a taste for single-malt whisky – the favourite is Glenmorangie – not all of them are totally integrated. Ted Brocklebank, one of Aberdeen's shrewdest journalists, told me of a hard-drinking evening he once spent with an American oil man: 'Late on I asked him how he was enjoying living here in Aberdeen. He was very drunk by that time and he said, "Hey, is that where we are?" It was just another oil town to him. He had his American company and his American store, his kids went to the American school, he talked on the phone to the guys back in Texas every day. He could have been anywhere.'

There was only one American aboard *Treasure Finder*, the meteorological officer, John MacDonald, a big man with a big belly and a big King Edward cigar stuck permanently in the corner of his mouth. His office was on the uppermost of the flotel's three main decks, next to the marine control department. He shared it with the radio operator and it seemed excessively small and cramped for the important work they had to do. The two men huddled in their respective corners like squirrels in their nests, surrounded by gear and paperwork and postcards from far-off places. The air was thick with the smell of MacDonald's cheap cigars; the atmosphere was casual and subdued. Voices squawked over the radio, the recording instruments hummed and clicked. There were large windows with a view of the Brent platforms and the bleak sea.

Despite appearances, MacDonald is unlike most of the Americans in Aberdeen. He is not an oil gypsy, temporarily settled in Britain, ready to move on to wherever the action is next – Brunei or Alaska or the China Seas. He is a man with a passion and is there by choice. Or rather, he is a man with two passions: for his tricky work and for the British Isles. He was born in Boston, Massachusetts, but his family scattered over the years and he joined the US Air Force in lieu of a better home. In 1956, the Air Force posted him to England, where he fell in love twice over: with an English nurse, whom he married, and with the place itself. In manner, he remains very American – relaxed, friendly, unbuttoned, a creature from a different world from most of his colleagues whose style is generally a little formal, withheld, and Royal Navy, as though beneath the tatty work clothes and turtlenecks there was always a crisp shirt and carefully knotted tie. MacDonald, too, is reticent, sober – he is allergic to alcohol – and less keen to talk about himself than about the arcane science of weather forecasting. Yet when I asked him why he was working in this remote place his large face softened and he stared out through the window of his office over the unfriendly sea with the eye of love, like a man who has come home. He said that after three years in England his unit was posted back to the United States. That Christmas, at an air base in Oklahoma, he had watched a television spectacular about Christmas around the world. 'Fifteen minutes of it was about England,' he said. 'And I sat there watching it with big tears rolling down my cheeks. In front of all my buddies. It was the only time I had ever felt homesick in my life and it wasn't even for my own country.' That memory stayed with him and when MacDonald retired from the Air Force in 1977, he returned to England, bought a house near Banbury, and began working offshore. He now commutes fortnightly between Oxfordshire and *Treasure Finder*. 'So he found a job in the East Shetland Basin and lived happily ever after' is an unlikely ending for any story. In John MacDonald's case, it is true.

Arthur Hewitt and Mike McCleave, the two men who run *Treasure Finder*'s administration office, put in longer hours than anyone else on board. 'We're up at five and sometimes we don't finish until midnight,' said Hewitt, a small man with an undernourished moustache and an incipient pot belly. 'We average around seventy hours overtime a trip. That's about two hundred and forty hours a fortnight. A seventeen-hour day.' Hewitt likes to break situations down into figures, like a professional poker player reeling off the odds and the outs. It is a habit of mind that goes with the job. He showed me a telex sheet like the one I had seen in the office of Ian McKnight, the OIS. It was about as long as a Rolls-Royce and listed afresh each day the names of everyone on board, categorising them by company and department. *Treasure Finder* has a permanent crew of one hundred; the other four hundred transients have to be shuttled continually around the field as they are required. It is work that demands great organisational skill, although Hewitt takes it as a matter of course. 'Up here nearly everyone's ex-service,' he said. 'So we're used to moving people around in large numbers efficiently.'

Mike McCleave is tall, thin, and lipless, with straggly grey hair and a gaunt face. 'There's nothing to it nowadays,' he said. 'It's just a matter of getting a routine and sticking to it.

But back in '77 and '78, when the field was being built up, we had to cope with thirty flights a day from Sumburgh. I didn't know what tired was until then. My home's south of London and I used to sleep all the way back to town – still do, in fact. Once I sleepwalked on to the wrong train and woke up in Brighton. The missus didn't take it too kindly.'

Despite the killing hours Hewitt and McCleave put in, their affability never falters. They have the specifically British talent for making the best of a bad job, the knowing, slightly down-trodden cheerfulness that got Londoners through the Blitz and used to be immortalised in Donald McGill postcards of gigantic matrons with their scrawny, lecherous husbands, and in souvenirs from Blackpool inscribed with poker-work aphorisms like ' "Cheer up," they said. "Things could be worse." So I did and they were.'

'Two trips offshore and you qualify as a North Sea tiger,' said Hewitt. 'It doesn't matter what job you do, a tiger's a tiger back on the beach. Stewards, cooks, dogsbodies who don't even smell fresh air the whole time they're out here, get back to Aberdeen, pull on their cowboy boots and swagger into the Yardarm, by Victoria Dock, saying they're toolpushers and mud engineers. Sometimes they'll even find a bird who'll believe them. No wonder people say North Sea marriages are doomed.' While he talked, he scanned one of his lists, checking off the names of the men waiting patiently in their clumsy survival suits for the next bus.

McCleave said, 'I was talking to a pilot yesterday. He thought . . .'

'The moment pilots start thinking everything goes up the spout,' said Hewitt authoritatively. 'They're just taxi drivers.'

On the last day of my first trip offshore I flew to Cormorant A, the most westerly of Shell's northern installations, to see Brent Log (short for Logistics), the flight control unit for the whole area. The previous day Brent Log's home base had been shut down while the drilling crew on Cormorant A fired an electrically controlled explosion into the oil-bearing strata far below. Since there is always a remote chance that stray radio waves might set off the charge while it is at the surface, radio silence had to be maintained on the installation and Brent Log had been moved temporarily to a makeshift office on Brent Charlie. Although it was a routine precaution, at that time it was being taken particularly seriously because a gas explosion on Cormorant A two weeks earlier had killed three workmen. Pinned to the notice board on *Treasure Finder* was an appeal on behalf of the dead men's families.

It was a brisk, sparkling March day. The sun shone, the busy sea was an impenetrable blue flecked with white, the horizon seemed limitless. The platforms lay spread across the shining water, their plumes of flame and smoke streaming cheerfully into the clear sky. We flew over Amoco's North West Hutton, brand new and painted bright orange. The Uncle John diving rig was moored near it, low in the water and very small in comparison. To the north, on the edge of everything, BP's Magnus was a vague shape piercing the rim of the horizon. Its loneliness made the cluster of Brent installations seem downright comforting.

Cormorant A is a huge four-legged concrete-gravity structure, held in place on the sea floor by its own weight, like Brent Delta. We circled it once in preparation for landing, then swung suddenly back out to sea. Over the intercom the pilot announced, 'We have a slight problem here. Cormorant Alpha is under red alert, so we can't land for the moment.' His voice was without expression, although a red alert indicates a high level of potential danger.

Warning lights are in place throughout each installation – in the cabins, corridors, work areas, recreation rooms, and mess halls – and each new arrival in the field is given a little card explaining their significance. A steady green light means the platform is working normally. A flashing yellow light means 'Stop all burning/welding and work requiring permits. Switch off all welding machines. Complete all crane lifts and set down loads. Evacuate columns. Await OIM's instructions.' When the red light begins to flash, 'Stop all work. Go to assembly points or duty hazard points.' The final stage is a flashing blue light accompanied by a howling siren, meaning,

'Remain at (or go to) assembly point. Await OIM's instructions.' The OIM's instructions at this stage are usually to get into the lifeboats and prepare to abandon the platform.

A red alert, then, was the real thing. I tried to picture the hullabaloo, men running to assembly points, struggling into their life jackets, orders barked over the intercom. It would be like a Hollywood disaster movie: *North Sea Inferno*. 'You've got to remember that if something goes wrong in an oil refinery you can run,' an oil man had told me in London. 'But if something goes wrong on an offshore platform, all you can do is jump over the side.'

In the helicopter, the man strapped in next to me said in a bored voice, 'Here we go again.' The man strapped in on my other side said, 'Welcome to Cormorant Alpha.' Nobody else took any notice.

The helicopter slid sideways until we were a mile or so from the platform, then began to circle patiently. The sun shone; the gas flare blazed peacefully into the mild sky. We were too far off to see any signs of activity on board. After five minutes, the pilot said in the same flat voice, 'We are back to green.' Two minutes later we had landed. Everyone on the installation was moving quietly about his work. There was no sign of an emergency and no mention of it later. The situation, in fact, was so blandly normal that, once on board, I forgot to ask why there had been a red alert. And even if I had remembered, the question, in that unruffled calm, would not have been appropriate.

Because Brent Log is the third busiest air traffic control unit in Britain, I had expected something large and shiny and sophisticated. What I found was a cluttered room not much bigger than a student's bed-sitter, and not all of it devoted to air traffic control. On the right of the door was a bathroom-sized cubicle, partitioned off from the rest of the room, housing Shell Marine Control, which coordinates the movements of all the shipping in the area – supply boats, support ships, tankers.

Brent Log itself was a battery of tired-looking radio equipment lining a workbench that ran the width of the room. At the right-hand end of the bench sat the two air logistics coordinators. One coordinates the crew-change aircraft to and from the beach – the Chinooks from Aberdeen, the Sikorsky S-61s from Sumburgh – organising the passengers, calculating baggage weights, passing on arrival times so that the helidecks are manned and ready. The other man does the same for the in-field flights of the Bell 212s, moving freight from platform to platform and positioning people to fill the vacant seats.

At the centre of the bench was a small Apple computer, its display screen grey and empty. When the appropriate data are fed into it, it will give the precise hourly position of any object in the sea – an oil slick, a buoy, a lifeboat, a man overboard. It is a piece of equipment everyone in Brent Log hopes to use as little as possible.

At the left-hand end of the bench sat the two men who control all the air traffic in the East Shetland Basin. Both were middle-aged, relaxed, and confident; both learned their skills in the RAF and now worked for a company called IAL (International Aeradio Limited), which contracts them to Shell. Well, yes, they said, it's still pretty busy up here – anywhere between three and five hundred movements a day. But some of those flights are very short – sometimes as little as a couple of minutes – and nearly all of them need nothing more than radio-telephone calls, not the elaborate talking-down process a jumbo jet requires when it lands. No problem, they said. Not like it used to be. They seemed almost nostalgic for the frantic first year, 1977–78, when their ramshackle little office was the third-busiest air traffic control unit in the world, handling twenty-one thousand movements a month. The Chinooks, they told me, flying direct from Aberdeen with their big payloads, had made their lives much easier. Yet despite the Chinooks and the levelling off of the size of the workforce now the whole field was operational, they had still logged a total of 849,526 air movements between their start in the spring of 1977 and the end of the previous month, February 1983. They expected to clock up their millionth movement well before the end of the year. 'It's the sort of thing we ought to celebrate,' one of them said. 'But how do you celebrate on a dry platform?' 'And who's got the time?' said the other.

The atmosphere in Brent Log was remarkably good tempered, despite the ceaseless radio chatter. Each exchange between pilot and air control required that the helicopter's position be located on a chart and then checked off against a board on which was listed the flight number, the names of the passengers, the nature of the freight, and the stopping places around the field. As each flight was completed, it was crossed off the board. The ATCs consulted their board, scribbled notes, and coped politely with my questions, constantly breaking off in midsentence to answer a squawking voice from one or the other of the loudspeakers on the bench in front of them. At their backs, a shoulder-high, glass-fronted tape recorder switched itself on and off automatically, activated by the distorted voices from the loudspeakers.

Behind the tape recorder was a row of steel filing cabinets that formed a half wall around the desk and chair that Tom Ogston, the supervisor of Brent Log, calls his office. Ogston's official title is offshore flying coordinator of Brent Logistics – otherwise known as Flyco, Brent Log – and he belongs properly to neither end of the office's workbench, being by trade neither an air traffic controller nor an air logistics coordinator. His particular qualification is a lifetime's experience with aeroplanes, both fixed-wing and helicopters. 'I know what I can ask my contractors to do and what not to do,' he said. 'At the same time, I can make sure that Shell is getting good value out of the aircraft in the field – the optimum use without overstretching them. And then, with my air logistics team, I can oversee just what the requirements are on every job. Let's say a toolpusher wants to move a drilling bit from North Cormorant to Dunlin; I know what it weighs and I also know that the floor loading of the 212 is about one hundred pounds per square foot; therefore we need to put in spreader boards and use a special technique of loading and unloading. I'm using all this knowledge of mine on behalf of the company and there's a lot of enjoyment in that, just by itself. Then there's the added pleasure of working with a good bunch of boys.'

Ogston's 'boys' looked to be in their forties, and he himself was fifty-seven – the oldest man I met offshore – although his thick fair hair was untouched by grey, his greenish eyes were bright and friendly, and he had the energy of a man half his age. But energy is an essential qualification for the job, since his day starts at five a.m. and does not end until the last flights have landed and the next day's flying programme has been worked out in every detail and telexed back to the operators in Aberdeen and Sumburgh and on *Treasure Finder*. Even when that is all done, Ogston and his team may still be called out during the night if a man is injured and has to be flown to hospital.

Ogston learned his flying not in the RAF but in the Army Air Corps, which used gliders – in the days before the techniques of the air drop were fully developed – to fly men and matériel behind enemy lines. When the Glider Pilot Regiment was disbanded in 1948 he switched to fixed-wing aircraft and spent the next eleven years flying little Auster spotter planes. In 1959 he transferred to helicopters, which he flew for another sixteen years, the last twelve as a training captain, travelling around the world initiating student pilots into the mysteries of the machines. 'A lot of the pilots out here – now working for British Airways, Bristow's, North Scottish – were my students,' he said. 'They all greet me warmly, which is nice. If I run into them back on the beach, they always buy me a beer. So I can't have been too bad.'

For twenty years, flying was Ogston's whole life. 'I was married to a career,' he said. 'At weekends I'd sit around the base and do air tests while the other chaps went home to their wives. I was totally happy: I loved planes, I loved flying, there was nothing else I wanted. But you change, you know, without even realising it. When I got to be about forty I began to think there might be something in this marriage lark. I also thought, I'm getting on; soon it'll be too late to start a family. It just so happened that that was when I met my wife.'

I said it must be difficult to be married to a man who not only flies unstable contraptions like helicopters but is also away much of the time. Ogston was unmoved. 'She's a service wife,' he said flatly. 'She was a nurse in the Queen's Royal Army Nursing Corps, so she and I have the same outlook.

When I pack my bags and tip my hat and say, "I'm off to Malaya, love," she just gives me a kiss and waves goodbye. She knows what the job's about. Nowadays, of course, I spend two weeks out of every four at home, so we see more of each other than we ever did before.' He paused, then added thoughtfully, 'Marrying added another dimension to my life; it didn't alter my love for the service.'

I asked how they met. He answered, 'It was a Good Friday and nearly everyone had knocked off for the Easter holidays. I was living in the mess and because my parents were dead and I was an only child I had no relations to go to. So I thought I might as well do some air tests. That involves two stages: first, the airframe and its ancillaries, then the engine. For both of them the aircraft has to be at maximum takeoff weight if the figures are to make any sense. When we loaded her up with fuel and crew we found we were about two hundred pounds short. So I said, "Ring up the hospital and ask one of the nurses to come." And this little girl arrived – twenty-one years old, four foot eleven and slim as a whisker. The first thing I said to her was "How much do you weigh, love?" "Ninety-six pounds," she answered. So I said to my chief mechanic, "Three more toolboxes, Fred," and off we went. It was a nice day, one of those soft spring days you get in the south of England, and she'd never been up in any kind of aircraft before. In an air test you have to fly very accurately and steadily, so I told her not to worry, I wasn't going to do anything desperate. And she loved it. When we got back down I asked her to come the next day while I did the high-altitude test. That entails a rapid climb, shouting out what the figures read at each thousand-foot mark. Well, Easter Saturday was another lovely, sunny day and off we roared to ten thousand feet. At that point in the test you have time to chat, so I pointed out the landmarks to her: there's Oxford, that's the edge of London, and down to the south you can see the whole of the Isle of Wight. I could see she was enjoying herself, so I asked her out to dinner that evening. We went to a nice little place between Amesbury and Salisbury. Candlelight and soft music. And that was it. Bang! Now here I am at fifty-seven with children of fourteen and eleven, while all my contemporaries' kids have left home and are earning money to keep Mum and Dad.'

He made it sound very simple. Like John MacDonald, Ogston seemed to be without malice, at peace with himself, absorbed in his work and curiously lacking in that skin of indifference that most of us acquire as we grow older, as though he were moving to a less broken and devious rhythm than that which impels people back on the beach. Listening to his youthful, enthusiastic voice, it occurred to me that what had brought me out to the North Sea was not just curiosity about the technological miracle in an impossible environment that had saved Britain from the economic stagnation that had overtaken Spain, that other decayed imperial power. I had also made the trip because I am a man with a sedentary middle-class occupation who wanted, just once, to be in a place where people deal with things other than pieces of paper and can see and quantify the results of their work.

With someone like Ogston, the rewards were obvious: quite simply, he seemed happy. But the other side of the coin was equally plain: the claustrophobia, strain, loneliness, exhaustion and boredom in the field; the drinking and broken marriages back on the beach. 'When you first come out it's a novelty,' said a technician who had worked offshore for five years. 'You fly on helicopters, you see things you've never seen before. But once you've been out here awhile it's like being in prison. You're cut off from all the normal things – trees, grass, people you don't know. You never get a news programme or an up-to-date paper. You see the same faces, day in, day out, for two weeks on end, and you get fed up with them, however decent they are. That can cause psychological problems. People get short-tempered for no apparent reason. You have to tread carefully; you can never really relax. It takes a special temperament to last out here and if you haven't got that you couldn't stick offshore life, no matter how much money you earned. I don't think anyone really enjoys coming out here to work. Nobody would do it as a career, or admit to themselves they're going to be working in the North Sea for twenty years.

The majority of the Shell guys are looking for promotion to an office job in Aberdeen. As for the rest of us, the idea is to acquire a bit of money – as far as the tax system allows – and start up businesses of our own.'

The man who said this was disgruntled but by no means alone. He had been sitting with three friends in the Brent Alpha mess when I spoke to him and whenever he hesitated the others urged him on. All four had one characteristic in common: they were neither professional oil men nor had they been in the services; all were civilians with specific specialised skills that were needed offshore. Later, on the same installation, I talked to a burly, bearded man called Nick Smyth who told me at length about what, for him, was wrong with life in the North Sea. Smyth had come offshore because he and his wife needed money for their farm in Wales where they raised sheep, ran a market garden, sold dog food, and rented out holiday accommodation. 'We had a very specific family plan,' he said. 'To carry it out, we could either win the football pools, rob a bank, or find some other way of acquiring a certain amount of capital. We still haven't found a suitable bank, so here I am.' He had previously worked as a group electrical development engineer for the British Steel Corporation; the job offshore offered him three times the salary for a fraction of the responsibility. Like Ogston, Smyth was an enthusiast, bursting with energy, yet however much he busied himself in his spare time – he was teaching himself to touch-type, was writing about his experiences in the North Sea, and took photographs that he sold to magazines – he still felt constricted and frustrated. 'I find it very dilapidating, both mentally and physically,' he said. 'Mentally, because I'm used to having an interesting job and now I haven't got one; physically, because it's impossible to get enough exercise. I walk around the platform two or three times a day. Big deal. I'd rather be a scaffolder, out in the elements. As it is, I feel I'm going to the dogs in every way. At home, if I want to be warm, I've got to get into the Land Rover, go up the mountain, find some wood, chop it into eight-foot lengths, load up a ton, bring it down, saw it up, and stack it in the woodshed three months before it's needed. Here, not only are you heated, you're too hot. On the farm, if I want an egg I have to go and shift one of the ducks or drive up to the village if the ducks aren't laying. Here, you just reach out and it's there. You get terribly lazy, and that's just one part of the mental degradation. Your brain goes to sleep. I think I have an APU – an auxiliary power unit – at the back of my head which does me offshore. But when I get back into the real world I have to start thinking again. My problem is that out here I just do not have *any* job satisfaction. When I was working for BSC I could look at a steel mill and say, "I designed and commissioned all the electrics for that forty-million-pound contract; and it works." When I came offshore I applied for a post as a senior authorised person – the fellow who issues permits to people telling them exactly what they can and can't do with high-voltage switch gear. It was a job I'd done many times and I thought, Great, I'm going to do some commissioning, I'm actually going to make things work. But the week I got out to the platform the organisation changed. Now all I do is run bloody cables about the installation, and I'm afraid I just can't get orgiastic about running cables.'

Back on the beach, Shell has a bureaucracy as large and intricate as the civil service of a medium-sized nation, and with all the usual civil service weaknesses: interdepartmental bickering, power plays, red tape, committee men cutting one another's throats. But out in the North Sea, where the action is, the bureaucratic jostling is kept in check by the need to get results. Smyth, however, had been caught in some kind of organisational finessing and had ended up with a job for which he was grossly overqualified. Hence his impatience, his frustration. He moved his hands as he talked, stretched his legs, leaned forwards and suddenly back; his voice was puzzled and indignant. It was like hearing a car with a faulty clutch – the engine revving frantically and nothing happening.

To add to his dissatisfaction, he missed his family. Ogston was grateful to his job because it allowed him more time with his wife and children than he had ever enjoyed in the services. For civilians like Smyth, the unalterable onshore-offshore routine was yet another deprivation. 'My kids are growing up and they're only kids once,' he said. 'I always seem to be away for the important occasions. Either you miss the run-up to Christmas or you miss Christmas or you miss the New Year. And whichever it is, knowing it's going to happen casts a blight on the festivities. I'm just not prepared to put up with it anymore – all the negotiation and sidestepping and body swerving just to get around the Christmas–New Year epic. It's not worth it. So at the end of the year I'm going home. I'll get some sort of menial job – sweeping the roads, anything that will bring in a regular income – and try to earn the rest from selling photographs and articles, from the holiday accommodation and the sheep. We'll have to economise but it will be worth it.'

'Then what?' I asked.

And briefly, the whirring engine faltered, his face clouded, his eyes became vague. 'Then we start something,' he said. 'The thing being unspecified at this moment. Whatever offers itself. But something we can run entirely from home – living at home, working at home. God knows how we'll manage. My wife does all the finances. In fact, there isn't very much she doesn't do, while I sit out here feeling useless. But until I say, "I've had enough, lass. I'm coming home and somehow or other we're going to find enough money to carry on," I don't think we shall ever succeed in anything. You can plan up to a certain point, but if you've only got twelve days out of twenty-eight to catch up, nothing is going to work.' Although his confidence seemed to return as he talked, his expression was hurt, as though 'Then what?' was not a fair question to ask an unhappy man. Offshore was the lion in his path and home the place where all problems would be magically resolved. What would happen after he disposed of the lion was not a question he could even begin to consider.

He opened a folder and brought out a set of photographs he had taken during one of those hundred-year storms: waves thundering around the legs of a platform, so high they seemed about to rip away the girders below the lowest deck, the whole ocean in uproar, the lights steady on the superstructure.

The camera he had used was cheap and elderly but the results were brilliant. I had seen a blown-up poster of one of them in Shell-Mex House in London and another version in a glossy Esso publication called the *North Sea Success Story*. It seemed strange that Shell should be able to use to the full the expertise and experience of a man like Tom Ogston, while all Smyth's talent and energy were allowed to whir away to no good purpose. Perhaps the reason is that the exorbitant expense of the North Sea installations is justified solely by the amount of oil they produce. They are a simplified world where nothing applies except work. Those who come out there are required to leave behind not only their families but also whichever areas of their personalities will interfere with the smooth functioning of this dedicated, oil-producing machine. The only way to share a cabin with another man, or three other men, for two exhausting weeks is to keep yourself to yourself, avoiding ups and avoiding downs, for the sake of peace and quiet and efficiency. In an environment as hostile as that of the North Sea the virtues that matter most are friendliness, good humour, and the kind of resignation that is learned in the armed services: a willingness to obey orders, however pointless they seem, and to accept the hierarchy. Imagination for anything except the job in hand is as great a handicap as excessive aggressiveness or a thin skin. During the time you are offshore, you are defined solely in terms of your work, and if the work does not satisfy you there are no other compensations. Smyth had hung on to his imagination but, because of his trivial job, had lost his self-esteem. As a result, his imagination and energy had turned in on themselves, magnifying his unhappiness. He seemed like a man in a cage. Watching him shift his burly frame uneasily about in his chair reminded me of the answer I had been given back in Aberdeen when I asked someone what would happen to the installations when the oil ran out. The man had laughed drily and said, 'Maybe they'll turn them into prisons.'

When I questioned John Hopson, Shell's head of aircraft services, about flying helicopters, he spoke willingly about the fascination of flying but could not remember that helicopters had ever given him any special problems. Tom Ogston, however, talked about flying, as he talked about everything else, passionately and enthusiastically, as though he had kept a hot line open to his youth. 'It's an entirely different technique,' he said. 'In a fixed-wing aircraft you're worried about stalling speed, whereas in a helicopter you're flying at no air speed or flying backwards or sideways. The coordination is different, too. A helicopter has extra controls; if you move one, you must move all the others. Each control has an effect that has to be countered by corrections on the other controls. Because a chopper is inherently unstable, you have to fly it with pinpoint accuracy and feel. In the first ten hours of training, your morale goes right down. You need a football pitch to hover the blooming thing. It feels sort of delicate and it takes you time to get used to the lag, to develop a sixth sense of when to put in a correcting movement and when to take it off again. It's mind-tying, like hopping up and down on one leg and trying to pat your head and rub your tummy at the same time. But after about ten hours, the penny drops. One morning you get into the chopper with your instructor and suddenly you've got it all at your fingertips. Then your morale soars. You can do things with a helicopter – put it in small spaces and work close to the ground – that make it far more enjoyable than a fixed-wing aircraft. In time, it becomes part of you; you wear it like a pair of trousers and can put it anywhere you like – literally, within inches when you're working with a winchman in an air-sea rescue operation. Of course, there are more principles of flight to learn, so in the classroom it seems more complicated than a fixed-wing aircraft. But that, too, adds to the pleasure because it's more of a challenge. All the time you're sitting in the cockpit – especially when you're hovering or doing delicate work with an underslung load – all those little red arrows that you've seen on the blackboard are working away in your brain. Anyone who saw you would say, "He's just sitting there doing nothing." But in fact you're thinking about those little thrust arrows and making tiny adjustments and balancing the machine like a juggler with a plate on a stick. It's the concentration that makes it interesting, and that's the secret of good flying.'

I had asked Ogston about flying because I was returning to Aberdeen on a Chinook and, courtesy of Captain Hopson, would once again sit in the jump seat between the two pilots. Before the aircraft arrived, I put my feet up for an hour in the cabin of one of the senior officers of Cormorant A, whose bedside reading seemed to encapsulate the preoccupations of life offshore: the previous week's editions of *The Economist* and *New Scientist*, six pin-up magazines, and, as a gesture to the lavish cooking, a much-thumbed copy of a book called *The Carbohydrate Counter*. I knew I had been away from home too long because the pin-ups looked good.

When I went up to the helideck the wind had dropped and the afternoon sun rode clear in a pale blue sky that merged imperceptibly into the ultramarine of the sea. In the distance, as it approached head-on, the Chinook was squat and bulbous, like a muscular, belligerent insect. 'Watch for the down-draught,' said the helideck officer sternly. 'It'll tear things right out of your pockets.'

Compared to the little Bells or even the Sikorsky, the Chinook was a huge machine. Its giant rotors fore and aft and two screaming jet engines in the tail created its own storm wind. Its great flat belly blotted out the sky. It seemed disproportionate to the helideck, to the installation itself, like a jumbo jet in a city square, although it touched down as delicately as a leaf. A line of windows – the same as those of a Boeing 727 – ran the length of its fuselage. Below them were the vast twin fuel pods that make the thing look so bulbous and also enable it to ride thirty-foot seas without capsizing. Inside it was like a modern airliner: comfortable seats upholstered in orange, luggage bins, a steward to welcome you aboard, and, when the doors were closed on the din outside, the faint sound of piped music. It seemed an unlikely machine to 'wear like a pair of trousers'.

It was now past three p.m. and neither of the pilots had eaten since breakfast, so someone was sent down to the platform's galley for food. While we waited, the chief pilot told

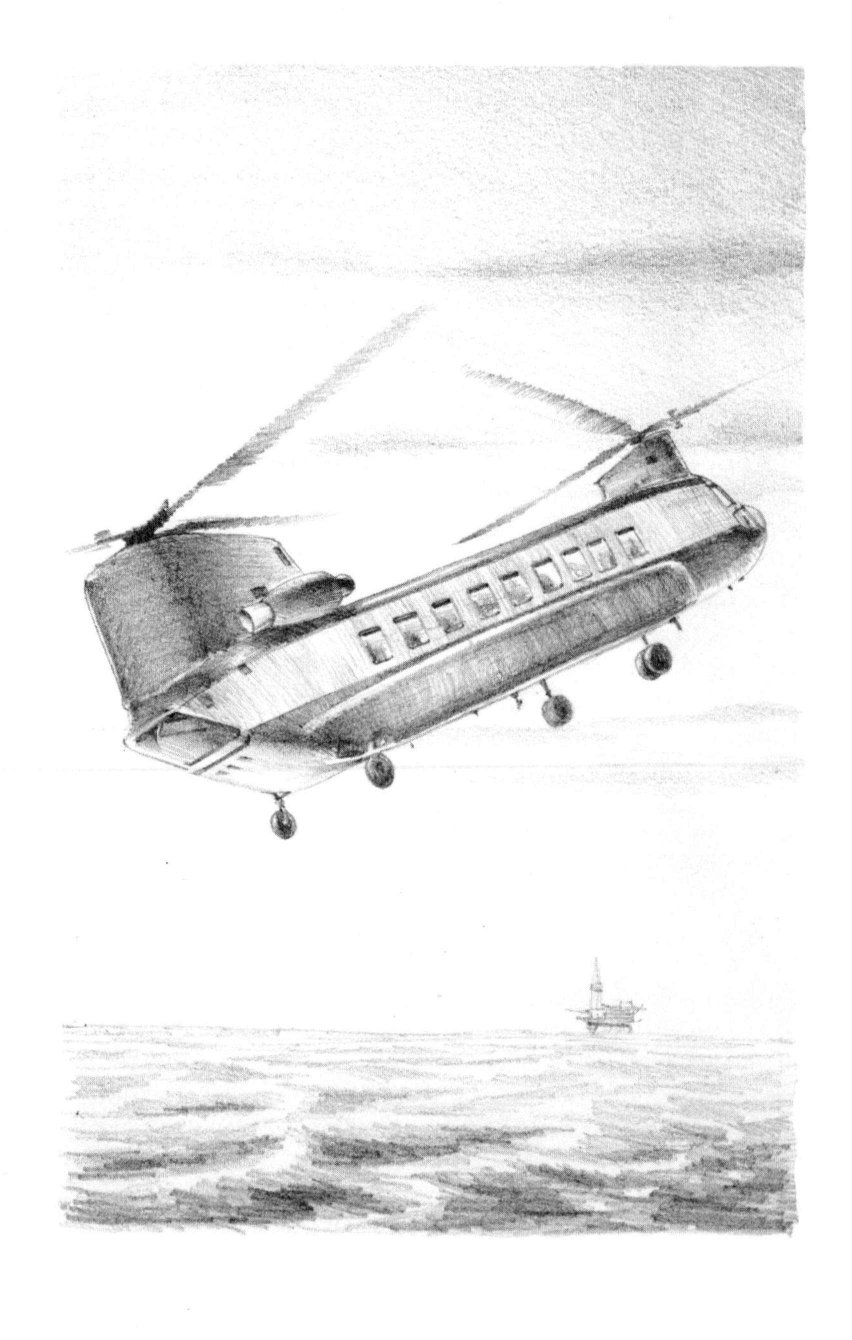

me, with considerable satisfaction, that they had just flown the three hundred miles from Aberdeen and held for seven minutes while in-field choppers were taking off; now they would return to Aberdeen without refuelling, and still arrive with enough fuel to hold, if necessary, for twenty minutes. 'Happiness,' he said, 'is a full fuel tank.'

The steward opened the door behind me and passed in a plastic tray covered with aluminium foil. 'At this moment,' said the second pilot, 'happiness is a full belly.' The chief pilot muttered something incomprehensible into his face microphone and listened intently to the reply. The men on the helideck ran for cover, the scream of the jet engines increased, the clatter of the rotors changed in intensity, and we rose effortlessly, very fast and very high. A great scroll of smoke rolled from the gas flare of a nearby installation. When we passed through it at three thousand feet it seemed like a cloud. I listened to the radio chatter between Brent Log and the pilots with the volume control of my headphones turned down low, while the sun blasted off the placid water far below. When we reached our cruising altitude of four thousand feet the chief pilot double-checked our compass heading, then locked the controls on to automatic and said, 'Lunch time.' The second pilot peeled back the foil from the plastic tray. On it were six weary little rolls with a single slice of processed cheese hanging limply out of each. The pilots eyed them with distaste. Finally, reluctantly, the chief pilot picked one up and bit into it. 'Yesterday's bread,' he announced. The second pilot said, 'Sonofabitch.' Even so, they patiently chewed their way through two rolls each, leaving the other two on the console between them.

Unlike smaller helicopters, the Chinook has a great deal of sophisticated equipment to make the pilots' lives easy: an onboard computer – called the flight direction system – that does most of the calculations that are normally done by the pilot and displays the results in a series of command bars which he then acts on; it also has an autopilot, like those used in commercial jets, and a new Decca weather radar. With this combination, a Chinook can follow a radio beacon and glide down a flight path to an airport automatically. When the first Chinook was delivered to British Airways Helicopters in July 1981, the director of flight operations, Captain Mike Evans, commented delightedly, 'The difference between the Boeing Vertol [the Chinook] and the Sikorsky we've been using can be explained like this: it's like going from a Rover 3500 to a Rolls-Royce.'

Yet on the flight back to Aberdeen, both the pilots looked tired, despite the labour-saving devices that enabled them to munch away on their dry cheese rolls with their hands off the controls. 'On a day like this it's child's play,' said the chief pilot. 'But we've had a rough spell. Each of these jets develops forty-five hundred shaft horsepower, while a small chopper makes do with seven hundred h.p. So fly one of these babies in bad weather for two or three days and you really feel you've been doing something.'

The radar screen at the centre of the control panel was large and technicoloured. Curved blue lines divided it into equal segments. When we took off, these indicated five, ten, and twenty-five miles, but the range was adjustable from two hundred miles down to two. 'On two we can pick out an oil barrel in the water,' said the chief pilot, and he juggled with it for my benefit. The green sweep moved backwards and forwards across the screen, backwards and forwards like a metronome, leaving a trail of glowing green dust that was the sea. Small objects came up in yellow; large objects were yellow with red centres, like ripe peaches. With the scale set on ten miles a yellow dot appeared at the top of the screen and moved slowly down it, growing in size. When the pilot switched to the two-mile scale the yellow dot became a large blob with a red core. Moments later, a supply boat appeared in the windows in front of our feet. 'Hey, presto!' said the chief pilot. Then he switched back to the ten-mile scale like a conjurer repeating a successful trick, while the sun blazed in through the right-hand window, baking us in our bulky survival suits. A northbound Chinook passed us five miles to the west, a fly crawling beneath a tail of cloud.

Just after five o'clock, two hours after takeoff, a big ragged mouth, flashing yellow and red, appeared on the fifty-mile band at the top of the radar screen. 'Land!' said the chief pilot. Despite the perfect conditions, he sounded relieved. The mouth grew rapidly as it crawled down the screen. The second pilot picked up one of the cheese rolls, peered at it critically, put it down again, and began to twiddle the controls on the radio. The steward appeared at the door behind us. He said that some of the passengers were booked on the last flight from Aberdeen to Glasgow at six-thirty and others had trains to catch; all of them were beginning to fret. The chief pilot checked the dials and the radar and sent back a reassuring message about our expected time of arrival. Moments later, the steward was back again for more reassurance. The closer the aircraft got to land, the edgier the passengers were becoming; after two weeks offshore, the idea of wasting a night in a hotel in Aberdeen was clearly not to be borne. The yellow and red stain crept slowly down the screen.

The chief pilot radioed ground control in Aberdeen about seats available on the Glasgow flight, then switched the radar to the twenty-five-mile scale. By five-twenty, the yellow mouth was nibbling the ten-mile band. Immediately, the pilot switched to the ten-mile scale, moving the yellow back to the top of the screen. It seemed to move now with a terrible slowness, although our air speed had not altered. Like the other passengers, I was impatient to be back. When the yellow stain crept below the five-mile line I craned forwards for a glimpse of the coast. But all I could see was a thick wall of cloud. The day darkened, the cabin became cooler as we flew into the murk.

The steward was back again, looking more harassed than ever. One of the passengers with a ticket for Glasgow did not yet have a confirmed seat. There was another flurried exchange of messages with Aberdeen Airport. They were sorry but the flight was fully booked. This was urgent, repeat urgent. They would see what they could do. Repeat urgent. They heard us and were trying. Silence. Seat confirmed. Thank you, repeat thank you. Roger and out. The chief pilot buzzed the steward with the good news and settled back in his seat. 'That's my bit done for Queen and country,' he said.

Ten minutes later, the cloud broke and there below us was the coastline: a long beach with its ribbon of breakers, dull fields beyond. All that green seemed strange after days of nothing but steel and water. The chief pilot switched the radar to the five-mile scale, buckled his safety belt, slid his seat forwards, and leaned towards the dials of the control panel, peering at them intently, as though seeing them for the first time. The second pilot pressed the switches to illuminate the NO SMOKING and FASTEN SEAT BELTS signs in the cabin. Both the pilots seemed suddenly tense and watchful, their gaze shifting continually from the dials to the landscape outside. Their concentration was like a physical presence in the cabin, almost palpable, although the aircraft was still flying on automatic. 'I'll bring her in on the radio beam,' said the chief pilot. 'So you can see the wonders of modern science.' We flew back into heavy cloud and stayed in it, although the altimeter read only 2,700 feet and we were dropping steadily at our 120-knot cruising speed. We flew inland, then swung south. Just below one thousand feet the cloud cleared at last. We were approaching Dyce Airport from the west, facing out to sea. The flight controller at the heliport said, 'Will you bring your speed up to one forty. It would help us. Thank you.' 'Look,' the pilot said to me. 'We're on the radio beam now.' He lifted his hands from the controls and the yellow band of the compass did not waver from its determined position. The runway approach blazed dead ahead of us: a long stem of white lights with horizontal branches at each side, just like a Christmas tree. At its top were two red and white Christmas candles, marking the beginning of the runway, then two lines of narrow white light that formed the edges of the runway itself.

At the last moment, the chief pilot switched over to manual. The great machine paused miraculously in its forward motion, as though gathering itself in, then dropped slowly and gently to the tarmac. We taxied across to the pretty white beehive that is the heliport terminal, scraps of paper whirling up around us. The pilot kept the rotors turning while a small bus bustled up for the Glasgow passengers. Lugging their duffel bags, they scurried across to it, crouching and at the double, staggering a little in the downdraught. Once the bus was clear, the pilot switched off the engines. The rotors spun slower and slower, drooping as though exhausted, then came to rest flopping downwards almost to the ground. I took off my headphones. After so much noise, the silence was uncanny.

It took only a few minutes for the bags to arrive in the customs shed but the delay seemed intolerably long. The men bunched together in silence, staring at the hatch, waiting for the baggage conveyor belt to start. When the duffel bags finally nosed through the hatch everyone pressed forwards at once, not wanting to waste a minute of time back on the blessed beach. The belt stopped, started again, stopped, started once more. Two pieces of luggage bumped slowly into the hall. A young man with short sandy hair and a tired face said, 'Last piece off again. Wouldn't you know it?'

An older man in a blue quilted jacket shouldered the other bag and said, 'Someone has to be last.'

'Why is it always me?' the young man said fretfully.

'I'll buy the first beer,' said the other man.

9

The granite city of Aberdeen has absorbed the oil business as easily as it absorbed all the other businesses that have kept it rich over the centuries: the fishing fleets and the whalers, shipbuilding and textiles, farming and cattle breeding, the quarrying and working of granite. As the largest city in northeast Scotland, it is the centre for the banks and law offices and insurance companies, for the chartered accountants and investment managers, all of whom went on flourishing as the cycles of industry changed. 'They couldna' have built these nice granite houses if someone hadna' been making money,' said Sir Maitland Mackie, who – when we met in 1983 – was Lord Lieutenant of Aberdeenshire. Sir Maitland himself lives in a particularly nice granite house in a secluded street of nice granite houses, not far from Rubislaw Quarry, which until 1971 was the source of Aberdeen's granite and is now a five-hundred-foot hole – reputed to be the deepest in Europe. Vegetation grows from the cracks in its slabs and buttresses, its bottom is flooded, and around its top is a high metal fence to discourage local climbers. Yet even though Rubislaw is no longer worked, granite remains the city's dominant building material, almost to the point of obsession: the little shed housing an electric generator, just outside my window in the Altens Skean Dhu Hotel, was faced by granite, and in the bedroom itself – not otherwise distinguishable from any motel room off any American freeway – the table was covered by a plastic sheet of granite look-alike. 'Aberdeen comes into its own in the rain,' said a melancholy, exiled Londoner. Yet paradoxically, this wintry grey city is filled in summer with millions of roses – red and pink and yellow and white – lolling their heavy heads in the parks and housing estates, even on the dividers of the ring road around the town. The Municipal Department of Development and Tourism claims that Aberdeen has won the 'Britain in Bloom' title so often that 'it now stands aside occasionally from the competition so as to encourage others!'

Granite and roses, dourness and wealth: the contrasts of the city are most apparent in the Old Town, where Aberdeen's famous, ancient university is situated. Its narrow, cobbled High Street is lined with low granite cottages interspersed with grand professorial mansions, also granite, behind wrought iron fences and barbered lawns. The university chapel, completed in 1505, has a square tower surmounted by a fretted stone crown; inside, the wooden pews are intricately carved with flowers and birds, oak leaves and acorns. But behind this peaceful façade is a big modern campus, all steel and glass and concrete, bustling and graceless.

The docks are the same: one side is efficient and shiny new; the other belongs serenely to another century. Somebody once said that Aberdeen earned its living from fish and chips, from the sea that it faces and from the rich agricultural land at its back. But fishing is now a far lesser concern. In the early 1970s, just as the oil men were moving in, a disagreement broke out between the in-shore fishermen and the 'lumpers', the fish market porters. As a result, the fishing fleets moved north to Peterhead and Fraserburgh, boycotting the port of Aberdeen. In Ted Brocklebank's words, 'It was as if God had looked down on Aberdeen and said, "Somehow we have to clear space in the harbour for the oil industry." ' Some fish is still brought down to Aberdeen to be sold in the market, but most of it is landed up the coast, and in the last decade Peterhead has become Britain's largest fishing port. The space cleared by the departure of the trawlermen was bulldozed and rebuilt to the most up-to-date specifications at a cost of somewhere between fifteen and twenty million pounds: new piers, new warehouses, new office blocks. The harbour is now crowded with steep-prowed, flat-sterned supply ships and ungainly seismic survey vessels, their superstructures seemingly top-heavy amidships, where the laboratories and recording instruments are housed. But behind the smart new docks and warehouses and brightly painted oil storage tanks is the nineteenth-century fishing community, Footdee, three enclosed squares of granite cottages that have walls two foot thick and tiny, precipitous staircases and glass panels engraved with schooners or flowers let into their narrow front doors. The houses present blank, windowless backs to both the harbour and the sea, and face inwards on to quadrangles like those of an Oxford college. At the centre of one of these quadrangles is the community's own minuscule chapel; down the middle of the others are lines of little wooden sheds in which the fishermen once kept their gear. In the days before running water, Footdee used its own wells and the massive iron standing pipes are still there, painted bright green, each with a white lion's head at its top, the creature's mouth forever gaping for water that no longer runs. In one of the squares is a ponderous clothes mangle, also freshly painted, its wooden rollers a matching green from damp and disuse. The quadrangles are old and beautiful and extraordinarily peaceful, considering the unceasing activity at their backs; they are also occupied mostly by the descendants of the people for whom they were originally built, being passed down like heirlooms from generation to generation, effectively excluding Aberdeen's Yuppies from moving in and dragging the place up market. On the seaward side of Footdee is a sea wall curved like a cornice of snow. Beyond it is a long beach of fine sand, broken by groynes and scoured by the east wind. Five hundred miles south, it would be a holiday resort and in deference to that illusion there is a ramshackle fun fair and a wide grass promenade, both utterly deserted on the bright, cold March day when I saw them.

When oil first arrived there were rumours elsewhere that Calvinist Aberdeen was becoming another Klondike. In reality, this meant that for a brief period men in ten-gallon hats and cowboy boots shivered in the dank hotels, the pubs were crammed with roughnecks and roustabouts, and a few canny locals made killings in real estate. But the American wildcatters moved in, made their lucky strikes, and promptly sold out to the major companies, often with the innocent stipulation that the fields they had discovered should continue to carry their wives' names. An exasperated American once said to Ted Brocklebank, 'You've got all those dumb broads out there in the North Sea – Beatrice, Thelma, Terri, Maureen, Beryl, Joanne, Josephine – all those wives of oil men.' But today nobody remembers who they are any more readily than they remember that Brent is the name of a goose.

In Aberdeen itself the effect of the oil industry has been equally discreet but very much more pervasive. Union Street, the city's main thoroughfare, has changed physically hardly at all in the last decade, but its shops are far better stocked than they used to be – the Marks & Spencer is one of the best outside London – and crowded with customers with money to spend. New hotels have been opened and old ones renovated, and there are now several ambitious restaurants of the kind that flourish on expense accounts. But the developments that have made Aberdeen into one of the richest cities in Britain – with the least unemployment and lowest suicide rate, and real estate prices as high as in London – are mostly out of sight in the new industrial estates at the edges of town where the Seven Sisters and the hundreds of companies that service them have their headquarters. 'We've tended to put the nasty stuff on the periphery of town,' said Brocklebank, although by British standards, the nasty stuff is not at all nasty and most of the oil companies' offices are a great deal more elegant and imposing than the blocks going up in London. Chevron, in particular, has risen to the place and the occasion with a magnificent Highland castle reborn into the twentieth century: grey granite, black slate, and tall pitched roofs to set off the acres of glass.

As for the cowboys: there are still plenty of Stetsons to be seen at the airport but their wearers seem to leave them behind in their hotel rooms. 'You could play "The Star-spangled Banner" on every corner in town,' an American oil man told me, 'and be standing there all by yourself.' Even so, at the Petroleum Club, out on the Braemar road where the Victorian merchants' houses are hidden behind stone walls and massed rhododendrons, the Americans have had a powerful effect on the tone of the place and the amenities. The atmosphere is cheerful and informal; members actually greet visitors instead of staring at them with the glacial dismay that is standard in London gentlemen's clubs. Behind the old main building is a brand new sports complex of pimply brick, with a swimming pool, squash courts, saunas, billiards room, restaurant, and a coffee shop looking down to the River Dee over an acre of car park and a golf driving range. Out on the lawn is a silver-painted barbecue big enough to handle a couple of Texas steers. It may not be as grand as the Petroleum Club in Houston, I was told, but it is the next best thing and nothing like as expensive.

Oil has not only brought foreign oil men and foreign money into Aberdeen, it has also brought back the native sons. For generations the Scots have been scattered across the globe like the Jews in the Diaspora – five million in Scotland, twenty million elsewhere. In Aberdeen in particular, according to Bill Adams of the Scottish Council for Development and Industry, 'There has always been a historic net emigration because the basic resources of the area have never been able to sustain the population. Although it has never been as drastic as in the Highlands or in Ireland – there have been no floods, no potato famine – the movement out has been steady because there was an inherent understanding that there wasn't a job for you when you came out of school or university. The eldest son inherited the farm or the boat or the business and the others had to carve their own way. This was taken for granted all through the education system. Young men knew that sooner or later they were going to have to catch a train to somewhere else, so they had to have a commodity to take with them, a skill. If they couldn't make it academically, then they made it through the crafts stream – they served their apprenticeship in the shipyards, say, and became marine engineers – because they knew at an early age that their destinies probably lay elsewhere.'

The oil industry has changed all that. In a nation debilitated by chronic unemployment, Aberdeen has become a place to find work. Native second sons no longer have to emigrate to Vancouver or Hong Kong or Valparaiso and the city's population has grown by thirty to thirty-five thousand in the last decade, not all of them visiting oil executives. Shell, for example, in collaboration with the local education authority, has started a training scheme aimed at catching local lads at sixteen or seventeen and turning them into oil field technicians. Britain has always been strong on technical invention, weak on business efficiency, and oil now provides an outlet for all that energy and cleverness, much of it the product of the School of Offshore Engineering at Aberdeen's distinguished science college, Robert Gordon's Institute of Technology. Bill Adams told me of one small company – a couple of dozen men, all graduates of RGIT – that produces underwater television cameras it now sells in places as far away as Australia and to atomic energy commissions as well as oil companies, since a camera robust enough to go down an oil well has no trouble in a nuclear reactor. 'To my knowledge, they have sent their managing director and sales director twice around the world in the last six months,' he said.

The coming of the oil industry also affected local business in subtler, less immediately calculable ways. Sir Maitland Mackie described what happened: 'We used to be competitive up here because we were so remote from other markets that we could operate on a low-wage economy. This kept employment up even during the Great Depression of the thirties. Then the oil boys came in and paid high wages for secretaries, for engineers, for chaps to cater for them on the rigs. And the local industries complained like hell: "How can we survive?" they said. But of course, they had to survive, which meant that they, too, had to start paying much higher wages. They therefore had to look to their efficiency. And that's been good for them because they have been forced to invest in modern labour-saving equipment. So the curious side effect of the oil industry's arrival in Aberdeen is that it has made local business and industry far more efficient. I hate to think what the economy of this area would have been like without the oil companies.'

Sir Maitland has a heavy, noble head, like a Roman marble, with deep lines running from the corners of his small mouth to his chin. His white moustache is carefully trimmed; his eyes behind his spectacles are bright and amused. He gets on especially well with the American community not only because he is clever and convivial – 'I canna' bear to count the numbers of bottles of my vintage port I poured down their throats before they decided to come here,' he said – but also because he has the priceless advantage of a wife from Texas. Apart from his many business interests, he owns a large farm on which he raises beef cattle. After the death of his first wife, he met, in America, a woman who was heir to a special breed of cows. As she tells the story, their decision to marry was really a decision to mix the herds. Lady Mackie is plump, shrewd, and startlingly informal for the wife of the Lord Lieutenant of the county. On the morning I went to their grand house in Rubislaw Den North she was wandering around in a battered housecoat, sleeves rolled up, hands wet and glistening with soap bubbles. 'We're gettin' ready for a tea party for blind children, and I've got no staff,' she said by way of apology. 'So I've been scrubbin' on garden furniture.' Sir Maitland is equally informal and direct but he also combines a canny flair for playing the government establishment for the benefit of the area with the ability to rouse local businessmen into the kind of go-getting activity that indolent British management either avoids or deplores. For example, directly the first vague rumours of oil in the North Sea began, he led a seventy-two-strong delegation of local businessmen and industrialists to Houston to discuss joint ventures with the Texas oil men. As a result of his charm and persuasiveness, a number of American companies – such as Baker Oil Tools, Halliburton, and Vetco

– have made Aberdeen the base for their operations not just in the North Sea and Europe but in the whole Eastern hemisphere, while the parent company back in the States runs the Western hemisphere. 'Being a wealthy farmer and landowner, with business interests in the hydrocarbon industry – he is a director of Highland Hydrocarbons, which tried to set up a plant in the Firth of Cromarty – and also being married into the Texas community, Maitland Mackie brings a unique breadth and dimension to the whole enterprise,' said Ted Brocklebank. 'He has a tremendous respect for the American community, and they for him.' In Sir Maitland's study, on a shelf behind his desk, sits one of the many tokens of that respect: a gaudily painted wooden statue of himself in a Stetson hat, with his knobby knees protruding between his cowboy boots and his kilt.

Sir Maitland's success is the success story of the whole area, but of the many individual success stories of those who saw their chance when oil came and grabbed it ferociously, the most spectacular is that of the late Robert B. Farquhar, whose company now builds, among other things, the topside structures for oil installations. Bob Farquhar was born in 1921 and left school when he was thirteen – 'Mi mither couldna' afford to feed me,' he said. He worked on a farm, where his wages were eight pounds for six months, then became a labourer for Wimpey, the construction firm. When the war came he joined the Royal Navy and was shipped out to India. After he was demobbed, he bought himself a horse and cart and travelled the countryside selling bags of firewood. With the money he made from that and his £65 gratuity from the Navy he bought a second hand truck. In postwar Britain there were shortages of everything and the whole country was in a state of chronic disrepair, so Farquhar began to diversify and expand; he built henhouses, sheds, garages, and greenhouses: 'Ye couldna' gae wrong,' he said. He set up a timberyard and by 1953 had won contracts to supply pit props to the National Coal Board and sleepers to the railways. Although the Beeching rationalisation of the railroads in 1963 shut down hundreds of miles of track, his business continued to prosper. It was a short step from sheds and garages to prefabricated houses – his company now owns a couple of holiday chalet parks – and from there to mobile living accommodation for industrial sites.

When the first rumours of oil in the North Sea began to circulate, Farquhar went down to Aberdeen's airport to watch the Americans arrive. He was impressed: 'If ye couldna' see 'em, ye could hear 'em,' he said. 'They were nae there for the fun of it.' He had turned fifty by then but was still as energetic and hungry as ever. He was one of Sir Maitland's first group that went to Houston to press the flesh and set up contacts, but his big break came, he said, when he won a contract from the local authority for the Grampian region to build a hundred mobile houses to accommodate the huge influx of oil workers moving into the area. Because his prices were competitive and he delivered on time, the oil companies took him up. He provided them with mobile homes, prefabricated office buildings, and site accommodation all over the northeast. When work on the Sullom Voe oil terminal was begun in the Shetlands he built accommodation units and also a complete air terminal building on the neighbouring island of Unst. In order to ferry the units there, he chartered a cargo ship large enough to carry whole houses. It was a long way from bags of firewood and a horse and cart.

Farquhar's efficient performance in the Shetlands made solid his reputation with the oil companies and he was ready for the next, most profitable step: the manufacture of prefabricated topside modules for offshore installations. He began by building living quarters and swiftly developed into more specialised constructions – kitchens, hospitals, dive-control cabins, laboratories, darkrooms, helicopter-control and radio rooms. His modules are now exported to offshore sites all over the globe – to the Middle and Far East, Alaska, Australia, and the Gulf of Mexico. His company also has major contracts with Shell, BP, Britoil, and Chevron for the maintenance and refurbishment of their existing offshore accommodation.

The Farquhar factory is in Huntly, a quiet little town in the rolling farming country northwest of Aberdeen, halfway along the road to Inverness. When I stopped at the edge of town to ask the way, I was told laconically, 'Ye canna' miss it. It's the big un.' Big perhaps, but not built for show. The main office is two storeys high and looks prefabricated; its corridors are narrow, its decor is strictly utilitarian. Bob Farquhar himself was also unlike any stock image of a multimillionaire. He was a small, Dickensian figure in a crumpled tweed suit: potbellied, red-faced, with big lips, jowls, two chins, and a full head of steel-grey hair. He shook my hand vigorously, eyed me rapidly up and down, and said, 'Let's gae u'steers an' ha' a noos.'

'Pardon?'

Farquhar beamed. 'Ay, a noos. A chat.'

Later, I gathered that his impenetrable brogue was renowned. When Sir Maitland Mackie visited a trade fair in Birmingham the local mayor begged him to go with him to the Farquhar stand to act as an interpreter. 'But the mayor spoke with a hell of an accent himself, so I had to interpret for both of them.' Sir Maitland sighed. 'I suspect Bob put it on a little.'

Upstairs was a long corridor lined with closed doors. Farquhar nodded at it: 'That's wheer mi laddies work – planners, surveyors, engineers, and sales chappies. Ay, an' we've e'en got one of those computer things.' He shuddered dramatically, then winked at me.

At the centre of the table in the boardroom, where we talked, was a delicate glass model of an oil rig under a glass dome. On a shelf in a corner was a painted scale model of a schooner – a present from his accountant, Farquhar said. But the rest of the room was bare and the surface of Farquhar's desk at the end of the conference table was uncluttered except for a telephone. I did not see the other offices but I had the impression that this room was Farquhar's one gesture to formality in the whole building.

As we talked, his brogue seemed gradually to clear. I wondered if Sir Maitland was right or if I was adjusting to the foreign language. Or maybe it was just his enthusiasm, which was no more intense about his multimillion-pound contracts with the oil companies than about the comparatively small businesses he had acquired along the way: a leisure centre in the Trossachs, the King George V Garden Centre, and the Gordon Arms Hotel in Huntly. 'I make a point of gaeing in there regularly. I watch the folk eat, then ask them if the meal's a' reet. Customers like a bit of attention. An' it keeps me going, you see. Ye ha' to keep going.'

While we were talking, one of his assistants called over the loudspeaker telephone to discuss a Japanese deal worth £1.76 million. Farquhar repeated the sum slowly, with no brogue at all: 'One point seven six, did ye say? Million. That's a tidy sum.' He shrugged and raised his eyebrows, the puir wee laddie astonished at his success. But when the assistant began to discuss the details of the deal, Farquhar switched over to the regular phone, answered in monosyllables, grunted, and made notes on a pad of paper he took from a desk drawer. After he hung up, he tapped the pad and said, 'I'm sixty-two and I'm beginning to forget. I used to keep it all locked up in mi heid, now I have to write things down. Age . . .' He gave me a withering look. He had had problems with his circulation, he said, and the doctors had told him that both his legs might have to be amputated. 'But I wouldna' believe them, so they gave me new veins and now I'm as guid as noo.'

He told me that in the early days his wife had worked as hard as he did, keeping the books, running things when he was away. Now they had two other directors for the company and a staff of two hundred and thirty. 'I think I've done mi bit,' he said. 'I'm gaein' to try to spend muir time wi' the wife. I used to leave at nine in the morning and she saw me again at ten at night. But now I've bought a house here in Huntly and I get haim for lunch. I want to sit back a wee bit but I always feel I need to be here looking over people. I've got guid management and it can run wi'oot me but I like to think it canna'. Unless you watch 'em, they spend money like it grows on trees. Still and a', I'm glad I'm not in Aberdeen. Oot here in a rural area there's nae problems with labour or unions. Country people are happy to do a job in these hard times and get haim to the wife and bairns. And I'll tell you something else: I drive a Rolls-Royce and naebody grudges me. When my first Rolls was delivered the men a' came oot of the factory and wished me health to drive it. They know I'm the one who's taking the risks.'

A more public recognition of the risks he had negotiated on the way to his Rolls-Royce and multimillion-pound turnover was a plaque on the wall naming him Scottish Industrialist of the Year, 1981. I asked why he had succeeded when so many other local businessmen had made fortunes out of the oil boom and promptly lost them again. 'I watched what I was aboot,' he answered. 'Them buggers in Aberdeen, their bowlers didna' fit their heids and doon they went.'

As doon we went to the lobby, I asked him if he really thought he would sit back now and take it easy.

'Wai . . . ill,' he drew the word out. 'In a coupla days I'm off to Nova Scotia and Newfoundland.' He winked conspiratorially. 'I keep mi ears to the groond, ye see.'

10

A Londoner's conception of the British Isles is bottom-heavy, like one of those nursery figures that can be rocked but never tipped over; Scotland seems like a distant, foreign province and even Edinburgh, 'the Athens of the North', feels as far off as Copenhagen and a good deal less cosmopolitan. But Shetland is genuinely the end of the line, even for the Scots. It is a remote, tightly knit community – seventeen thousand when the oil came – nearer to the Arctic Circle than to London, with tenuous links to the rest of the country and vague aspirations of independence – particularly from Scotland. Its closest connections are with Norway, where many of the early settlers, local customs, and place names originated. 'But Shetland isn't Norway or Scotland or Orkney,' said a pale young Shetlander called Stuart Robertson who works for BP at Sullom Voe. 'It has a steady independence, nothing wild. We have our own traditions and identity and we want to keep them. There was no television here until 1964, no BBC2 until 1977. And that was fine by us.' 'Shetlanders are a Nordic people, like the Icelanders,' said Ted Brocklebank. 'Scotland happens to be the nearest link they have to hospitals and suchlike amenities. But really, they are quite happy on their own.' 'It's a land of realities,' said Peter Guy, an ex-RAF officer who came to Shetland, fell in love with the place, bought himself a house on the island of Yell, and now works for BP. 'There is no room for flannel, for posturing. It's too exposed in every sense and too remote – eighteen hours by ship from the mainland, one hour by plane if the weather permits; and it often doesn't. Yet the people are seafarers, so they have seen the world. Most of my neighbours on Yell know the Falklands better than they know Edinburgh. And there are some old Shetlanders who have been to Hong Kong but have never visited the other parts of Shetland. Once they are home, they stay put. They are a self-sufficient people.'

Shetland maintained its independence thanks to three quietly flourishing industries – fishing, fish processing and knitwear. Then came the oil, turning the islands, first, into a staging post to the offshore fields, then into a landfall, a terminal where the oil was piped ashore, and bringing with it wealth and, in places, what Brocklebank called 'the unacceptable face of the oil industry'.

The mainland of Shetland is shaped like a jagged dagger pointing south, with a large pommel that slopes down from northeast to southwest. The blade is the narrow southern strip between Lerwick, the capital, and Sumburgh at the southernmost tip, a bleak, woebegone area made even bleaker by haphazard new development. Just north of Sumburgh Airport, the huge tropospheric scatter receivers that connect offshore radio traffic with the Aberdeen telephone system perch along a ridge of hills like invading Martians spying out the land. When I was there in July, the roadside was thick with buttercups and daisies and flowering white clover, and beyond the dry-stone walls were fields of yellow marsh irises and brilliant patches of poppies. But scattered randomly across them, like an impetigo, were drab little houses, most of them apparently prefabricated, all of them seemingly impermanent, although they were probably there to stay.

Seagulls squabbled for scraps in the middle of the road, then rose, angrily complaining, from in front of the hire car's wheels. Yet the coastline, even in the dingy south of the island, was extraordinary. A bay, an inlet, a headland, another bay. A coastline like fern. A reddish headland shone in the sun, a belt of purple mist across its middle. The cliffs in the distance were washed in blue. Then the road turned inland across moorland cropped by sheep and cut by peaty streams, before dropping down to Lerwick, row after featureless row of grey houses, becoming older but no less grey or featureless the closer they were to the centre of town and the harbour. At the dockside there was a hotel, a couple of pubs and cafés, a shop that sold videotapes, and acres of warehouses, some of them very large and very new. A roll-on, roll-off passenger ferry was tied up among the fishing boats. Out in the great, glittering bay half a dozen fish-factory ships lay at anchor – German, Norwegian, Russian – waiting to buy from the local trawlermen.

North of Lerwick, the road was empty and the countryside was what it had always been: rolling moors swept clean by the wind, sheep scattered across them, sheep on the road, little farms – called crofts – few and far between, villages fewer and farther. Once in a while, a truck thundered south on its way from the oil terminal at Sullom Voe. Apart from them, I saw no one for twenty miles until the road forked and I went right when I should have gone left. A couple of miles on, I saw a figure striding along the road who looked as I imagined every young Shetlander would look: ruddy complexion, innocent blue eyes, a shotgun over his shoulder, a dead rabbit dangling from his left hand. I pulled over and asked him the way to Hillswick, speaking slowly and clearly because of my English accent. He answered in purest, Bow-bells cockney: 'Sorry, squire. Yerl'avter gow back.'

A few miles further on, on a desolate stretch between Brae and Hillswick, with nothing but rolling moorland spotted with grey tarns and not a house in sight, I saw far off another figure marching purposefully along, carrying a suitcase. It was a young black in a New York Yankees bomber jacket; the suitcase was an outsized stereo recorder, playing full blast. He did not return my wave as I passed.

All these outsiders have been lured to the Shetlands by Europe's biggest industrial development in the 1970s, the multimillion-pound deep-water terminal that currently handles 1.2 million barrels of oil a day, half of Britain's daily

North Sea output, reputedly worth a gross one million pounds an hour to the producing companies. But to the ordinary visitor the terminal might hardly exist. It is tucked away on the far northeast edge of Sullom Voe, the great inlet that separates the dagger's pommel from its handle, a bleak, empty area with nothing to attract the fly-fishermen and ornithologists and amateur archaeologists who, apart from the oil men, are the only people who make the long journey to Shetland. 'Sullom Voe was one of our black spots. It had been an Air Force base during the war and was really run down. So the good Lord was kind to us when He put the deep water in a place that was already wasted. That's what you call forward planning!' So said Ian Clark, who was chief officer of the Zetland County Council when the oil companies decided to bring the oil ashore in Shetland.

If the Shetlanders believed in the cult of personality – and they vehemently do not – there would now be a statue to Clark in the centre of Lerwick, for it was he who, almost singlehandedly, transformed the arrival of oil in the islands from what might have been a social and environmental disaster into a source of great wealth for the whole community. Clark eventually became a managing director of joint ventures for Britoil, an oil company in which the government has a 49 percent share, and we met at Stornoway House, Britoil's stately London headquarters, just around the corner from St James's Park. A sleek, lone receptionist sat behind an elegant desk in the reception hall, which was paved in black and white marble; the atmosphere was hushed and calm. Not so Ian Clark, who is strong-voiced, purposeful, good-natured, and tough, like a warrior-monk on a medieval crusade. One of his front teeth hangs down in front of the others, making his smile seem jagged and oddly fierce. He laughs easily and often, and his laugh resonates through the deeply carpeted corridors like an underwater explosion. His large, comfortable stomach and shock of white hair make him seem, at first glance, twenty years older than he is – forty-four. But his face is not much lined, and even when sitting quietly behind a desk, he buzzes with youthful energy.

Clark is a Lowlander, not a Shetlander, born and raised in industrial Lanarkshire, just outside Glasgow. He left school at fifteen and went to work in a firm of chartered accountants. In his spare time he took a correspondence course to qualify as a certified accountant ('Britoil constantly boasts that I'm the only certified person here!' he said), then moved into local government. He spent fifteen months in Kirkcaldy as senior internal auditor, then five years in Berwickshire, where he rose to the rank of deputy treasurer. In 1968, he moved to Shetland: 'The attraction of Shetland to me was twofold,' he said. 'One, I reckoned that it might survive the reorganisation of local government that was being talked about at that time; two, in 1968 I was a mere twenty-nine-year-old and I knew that no other local authority was going to offer me a chief officer's post.'

In the four years between Clark's arrival in Shetland and the oil companies' decision, in 1972, to build an oil terminal there, he became a major figure in the community. He cajoled central government into funding a roll-on, roll-off vehicle ferry service to Lerwick – 'a dream they had had since the early fifties which had never materialised.' More important, the islanders trusted him: 'I was regarded as someone who was working for the community. I think it mattered that I have a very, very strong religious background. All my life, until I came back from Shetland, I was linked with the Plymouth Brethren, so the islanders knew me to be a committed Christian. And between my church activities, my reading, and my work, I didn't have a great deal of time or appetite for socialising. That meant that I was working away to get things done and was not seen by them to be linked with any particular grouping. I was regarded as being somehow above all that, a sort of pure white man. So when the pressures began to build and outside parties tried to stir things up, the community never lost faith in the fact that – whether I was right or wrong – I was working in their interests.' He hesitated, as though catching himself out in some immodesty. Then he added, 'There are two other things that worked to my advantage: when oil broke in 1972, I was thirty-three, so I had all the vigour of a young man to meet a very heavy burden. But I looked fifty-three. My hair turned white when I was in my twenties and I've always had a middle-aged tummy. So I was a sort of father figure to them.' His laugh erupted briefly and the faint sound of typing in the next room stopped for a moment as if disturbed by the shock waves.

Clark continued: 'I knew that if I was going to cope properly with the coming of oil it would no longer be a job but an all-consuming vocation. I suppose I could have gone to sleep and no one would have noticed, but that didn't seem right. I spent a lot of time in prayer. Finally I came across some verses in Paul's Epistle to the Ephesians: "Servants, be obedient to them that are your masters . . . but as the servants of Christ doing the will of God from the heart; with good will doing service, as to the Lord, and not to men." And I thought, Well, there you are; the greatest service I can do my Maker is in trying to serve the community in this way. So in all due modesty I prayed, "Lord, there's no way I can do this on my own. But I'm willing to try, so I'd better get some help from You!" ' Another explosion of laughter.

From that point on, Clark worked seven days a week. From Monday to Friday he was in his office from eight in the morning until six in the evening; then he went home, ate his dinner, and worked until ten. On Saturdays he worked at home from eight a.m. to one o'clock and from seven p.m. until nine-thirty. On Sundays he worked most afternoons and all evening. 'I told my wife it kept us from quarrelling.' The purpose of his labours was to protect Shetland from a threefold risk: to the environment, to the economy, and to the social fabric of the community. As he saw it, if the oil companies moved in on their own terms, a few people might make a lot of money in a short time but the islands would be spoilt, the community divided, and local industries would fail because their workers would be lured into high-paying temporary jobs in oil. He decided that the only way to prevent this was to join the oil companies as a partner, not a subordinate, sharing in both the cost of development and the profits from it, and to maintain control in a simple feudal manner by owning the land on which the development would take place. He also decided that the oil companies should contribute 'disturbance payments' – a kind of tax based on the tonnage of oil and gas exported from the terminal – that would go into a charitable trust to be used to relieve social hardship, support local industries, and act as a cushion against the time when the oil companies eventually moved out.

It was an audacious, improbable plan. Even national governments find it hard to cope adequately with the oil giants, and the Zetland County Council, as it was then called, was, in Clark's words, 'a sleepy, wee local authority, one of the smallest in the whole of Scotland.' But Clark himself was anything but sleepy; he was confident, clever, and utterly unfazed by the high-powered company he suddenly found himself keeping. This is how he described what happened: 'While the rest of Britain was going into recession, the Shetlands were knowing their greatest prosperity in this century. The fishing was good, the fish processing was good, the knitwear was good. Everything was coming together and the attitude was buoyant. So I said to the oil industry, "We don't need you and the community doesn't want you. But we recognise that if there is a lot of oil out there you will do your calculations and you will decide whether it's cheaper to take the oil direct to the mainland or to bring it ashore here. Even if we wanted you, if your calculations tell you to take it to the mainland you will do that and we couldn't attract you. And if we didn't want you, if your calculations tell you it's cheaper to use Shetland we will be pressurised into letting you come because the nation needs the oil. So we will not embark on an internal argument about whether or not we should have you. What we are going to discuss internally is this: *If we are unfortunate enough to have to welcome you, on what terms will you be welcome?* First," I said, "we are going to protect the environment, so you can forget all about your single-user terminals." That upset them. "Nowhere in the world do companies not have their own terminals," they said. "Nowhere in the world?" I answered. "If you want to use Shetland, it will be the first time in the history of the world that there will be a joint-user terminal. Second, when the oil is offshore or in the terminal the risk of pollution is not all that great. The danger comes when the vessels are being loaded. So we are going to be the Ports and Harbours Authority in order to oversee safety. Next, you can forget all about planning permissions, planning consents, etcetera. One would have to be a genius to foresee all the things that have to be provided against, so planning control is not going to be sufficient. Instead, we are going to take ownership of the land on which you build so that we can have an ongoing landlords' control. In fact, we are going to participate in the development in a form that gives us a direct say and a direct influence over it."

'At that time, I was talking about a fifty-fifty company to develop a terminal which they had costed at fifty million pounds. So I went down to the City of London and made sure that I could raise more than twenty-five million. (Actually, I could have raised sixty million.) Then I went back home and didn't reveal this to anyone. When Frank McFadzean, the chairman of Shell, flew up to Lerwick, he told me the fifty-fifty company was absurd; we Shetlanders would never be able to afford it. I answered, "Let me make you an offer: you tell me how much money I will need to raise and give me forty-eight hours. If I can raise it in that time, you take me on as an equal partner; if not, I'll forget all about it." But of course, he wouldn't look at it. He knew I would not say that without having taken out insurance. So they tried another tack. They invited me and Mr Blance, the Convenor of the Council, to a joint meeting of the companies at Shell Centre in London. They told us that our price was so high that they had decided they could not afford it and would not, after all, be coming to Shetland. I answered, "First of all, I want to give you our warm congratulations. I am very impressed that you have the technical ability to bring the crude direct to the mainland of Scotland on an economic basis. Second, I give you even warmer thanks and assure you that when the Convenor and I return to Shetland and announce that the oil industry has decided not to make its home there, we will appear as heroes. And third, if we can introduce you to any of our colleagues in local authorities in Scotland, we will be delighted to do so. Our relationship to date has not been the most harmonious and I consider it absolutely delightful that it is ending on this friendly note. So before we undo any of that, I think the Convenor and I should take our farewells of you. Please, gentlemen, be assured that you have our infinite gratitude. We thank you on behalf of the entire Shetland community." So we stood up to leave, but before we got to the door we were pushed back into our seats and told that, of course, they had got to come to Shetland.' A pause. A seismic explosion of laughter. 'At that point, they had rather disturbed the negotiating balance.'

As a result of that meeting, an Oil Liaison Committee was set up by the Council and the industry in 1974. In the same year, the Zetland County Council Act, designed by Clark to give him the necessary powers to control the whole development, was passed by Parliament. The act gave the Council everything Clark had wanted: the right to acquire land, to invest in commercial undertakings, to license construction, and to act as Port Authority at Sullom Voe. The Council would make land available for the terminal, provide harbour facilities, and build a village for the construction workers; in exchange, it would receive disturbance payments from the industry.

The disturbance payments, designed to provide for the future prosperity of the community after the oil had gone, provoked another confrontation in London that Clark also recounts with glee: 'Mr Blance and I were called down to see the Secretary of State for Scotland, who was at that time Willy Ross. And Willy Ross said that he had received complaints from the oil companies about us and he read us a wee lecture. So the Convenor had his say and Ross read us a further lecture. Then I said, "Secretary of State, I am very confused. My understanding has always been that the oil industry is the most powerful industry, if not the most powerful grouping, in the whole world, and the Shetland Islands Council is one of the smallest and weakest local authorities in the whole of Scotland. I had assumed that the Secretary of State for Scotland was responsible for local government. Now, had you been making these noises to the oil industry in order to protect the Shetland Islands Council from them, *that* I could have understood. But to the Shetland Islands Council on behalf of the oil industry!" And Ross laughed and said, "All I'm saying to you is be reasonable!" I answered, "Secretary of State, would you care to define reasonable." "Away back up to Lerwick, Mr Clark," he said. "You know full well what reasonable is. Just be reasonable." At that point I knew we had won the game: the industry had thought they were playing their ace and their ace had been trumped. So we negotiated the disturbance payments that have already brought tens of millions of pounds to the islands.'

One of the oil men most closely involved in the negotiations with Clark was John Heaney, who was technical director of Shell Expro at the time and the driving force behind the creation of the pipeline from Brent to Sullom Voe. Heaney is now in his middle fifties, a resolute man with a slightly squashed face and dark blue eyes that curve orientally when he narrows them. He was on two expeditions to the grim islands of the South Atlantic before he joined Shell, and subsequently worked for the company all over the globe. When I met him in 1983 he was dividing his time between fruit farming in Essex and running Saxon Oil, an independent company he helped to form in 1980 when he retired from Shell. He is efficient, hard, and fit – he walks with the air of someone who could walk unbothered all day – but even he seemed uneasy when he recalled the negotiations with Clark. I asked him how the Shetland Islands Council had managed to call all the shots, despite its weakness and the power of the Seven Sisters. 'It was not just one company in the negotiations, it was the whole industry,' he answered. 'What you must appreciate is that, while we were pioneering on all sorts of fronts, we felt that wherever possible we had to employ conventional, well-tried techniques. It was a question of prudence. You must minimise your risks whenever you can and only go in for the real frontier stuff when there is no proven alternative. Now, people understand pumping oil to shore and handling it in tank farms, as they are called. So the industry collectively felt they had to try their best to get the crude to Shetland. Perhaps it would have been different if it had been just one company head-to-head with Clark. But as an industry, we never called his bluff.'

It is characteristic of the Shetlanders that the first to benefit when the money began to flow were the exposed and under-privileged. The old-age pensioners, who had stuck it out when the going was hard, were allotted a special bonus each Christmas. The disabled were provided with hand-controlled Volvos to replace their unstable little National Health three-wheeled invalid cars that were deathtraps in the fierce winds that scour the islands. 'Using the money in that way drew the community together psychologically,' said Clark. After he left Shetland, more grandiose schemes were begun – Lerwick now has a three-million-pound sports complex, the largest building on the island – and even Clark was unable to prevent locals from being lured into oil-related jobs that paid three or four times the amounts they had previously earned. Since oil arrived in Shetland, the roads are better, the schools are better, the drainage is better, but life generally is more expensive and the indigenous industries have declined, despite all Clark's efforts.

Some people think there has also been a falling off in other, less tangible ways. Brian Lappin is an Esso man who was seconded to Shell Expro to become the first industry chairman of the Sullom Voe Association. Even now, a decade later, when he talks about his meetings with the islanders to discuss the landfall of North Sea oil and the construction of the terminal, his formal, boardroom circumlocution cannot quite disguise his distaste and dismay for all that stern horse trading: 'There wasn't really a high degree of community of viewpoint between these two groups of people who had to act together,' he said. 'I think it is also true to say that the oil companies didn't really anticipate with great skill or reasonableness the degree of resistance or restraint or control that the island community was expecting to exercise over the oil companies' activities. To say that the companies were high-handed is perhaps an unreasonable conclusion, but certainly the community and its representatives regarded them as high-handed almost by definition. That was really their starting point. So, yes, the early days were' – he cleared his throat – 'quite difficult. In the end, we worked it all out and gave them what they asked for to the best of our ability. Even so, not everything is perfect. Traditional Shetlanders insist that the outlook of people on the islands has been affected. Up until the early seventies, Shetland was so remote and the maintenance of life was such a hard task for everybody that there was a tremendous feeling of community: if a person or a family got into trouble, the others would rally round and take them through the bad time. It is put to me that now that sort of friendship with one's neighbours, which had its origins in real necessity, has been eroded. People have become more inward-looking and selfish, less caring about the interests of others.'

Like many of the rumours about the way oil has changed life in the Shetlands, none of this is immediately obvious to the visitor. For example, on the hill above the St Magnus Bay Hotel at Hillswick is a crafts shop that sells beautiful local knitwear and ugly little sealskin coin purses. The shop is the front room of a cottage owned by Ivimey McKenzie, a talkative old lady who uses two walking sticks to hobble around her house and can no longer negotiate the hill down to the village store. She told me she had come north from Glasgow because her daughter had married a Shetlander; when the young couple emigrated to Canada, Mrs McKenzie stayed on. Now she makes do for company with two small dogs, a sheltie and a *café au lait* terrier, with whom she carries on lengthy, animated conversations. But she was never lonely, she said: 'In summer I have my customers and they come in all shapes and sizes. The other day I had two naval officers in here. Weighed down with gold braid they were. But ye couldna' get a smile out of either of them and they didna' buy a thing. Then ye get the laddies from the oil companies, all trussed up in their business suits, puir dears.' But in a remote spot like Hillswick business is never brisk and in winter, when there are only four hours of daylight, Mrs McKenzie sometimes sees no customers for weeks on end. Her neighbours, however, look in on her all the time. They do her shopping for her – one was there with a basket of groceries when I arrived – and when that excuse is lacking they simply stop by for a cup of tea and a chat. 'They look after me,' said Mrs McKenzie. 'They look after each other. That's what they're like up here.'

The closeness between Shetlanders is based on concern, not on gossipy village curiosity. But it has a curious and inconvenient side effect: outside Lerwick, there are hardly any public houses. And the pub, in Britain, is an institution, not a shadowy place to get drunk in, like an American bar, but a social centre where the locals gather at the end of the day to chat over their beer. Pubs have darts teams and shove ha'penny teams that carry the village flag in competitions around the district. The pub is where the football and cricket teams pin up the list of players for their next match, and where village outings are organised, village news is disseminated. But not in Shetland. There was a lively pub in Hillswick, a low, stone building by the jetty, and I saw another in Brae. Apart from them, there seemed to be none in the villages and hamlets scattered around the mainland. I asked Ian Clark how such a closely knit society managed to survive without this most traditional of British community centres. He answered: 'The social fabric is so strong you don't need them. Most Shetland houses are as well stocked as pubs, and you know someone regards you as a close friend when he walks in without knocking while you are reading your newspaper in the evening, sits himself down, and takes a drink with you. It's a sign of friendship. Who needs pubs when you can wander in and out of each other's houses?'

As well as its pub, Hillswick has the St Magnus Bay Hotel, one of the two good hotels north of Lerwick. (The other is the Busta House, near Brae.) The St Magnus is a big, white-painted, wooden mansion that was built in Norway in the 1890s and shipped across to Shetland at the turn of the century. The interior is all creaking, varnished pine, like a house in a Chekhov play, and some of the staff have a matching Russian eccentricity. Des, the young headwaiter, was an intense bean-pole of an Englishman, whose father was mayor of Eastbourne and whose sister worked for the British Council in Belgrade.

When I asked him why he was in Shetland, he shrugged despondently and said, 'There's always work in the catering trade,' and explained that he had a Cordon Bleu diploma and had once worked on a project for the Roux brothers, London's most influential chefs. This seemed to deepen the mystery of what he was doing in Hillswick, although it explained why he fussed inordinately over the way the simple meals were presented and why his eyes lit up like a lover's when he talked about claret. He also talked, with equal passion, about archery, and, more dreamily, about the restaurant he was going to open once he had accumulated some capital. 'In the south somewhere,' he said wistfully, like Olga, in *The Three Sisters*, yearning for Moscow. 'Definitely in the south.' Shetland, apparently, was not part of his dream.

Unlike the hotel manager, a dark, bearded, diffident Glaswegian who had spent five years offshore working as a maintenance engineer, an experience that had earned him a lot of money and two nervous breakdowns. 'I'd work for three days at a stretch if something important broke,' he said. 'Three days, three nights without a break. I'd say to them, "Just let me have twelve hours' sleep and I'll finish it quicker." But they never could. Having machinery out of action costs too much. I cracked up twice and then I quit.' With the money he saved he bought himself a seventeenth-century Shetland croft with stone walls four feet thick and thirty acres of land. 'I love it here. It's another world and I never want to leave. My children go to the local school: forty kids in four classes; individual tuition, more or less. There's nothing like Scottish education. But . . .' His voice trailed away and he looked around restlessly, his nervousness seemingly undiminished by the idyll of this new life. 'But it gets very lonely at times, particularly if the snow comes and stays.' We were talking in the lobby. Behind him, framed in the glass doors of the hotel, two boats bobbed on the bright water just offshore. Beyond them was the great sweep of St Magnus Bay, ridge after blue ridge reaching out into the sea, the spray rising peacefully, far off, around their feet. The day was hot and cloudless. He made a gesture with his hand, outwards and upwards, as though to obliterate the scene. 'There are three things you need up here that might seem luxuries anywhere else,' he said. 'A car, a deep freeze, and a citizens band radio.' A pause. 'There's not much light in winter,' he said. 'You have to be able to keep in touch.'

'Shetland is a dream or a nightmare,' Ted Brocklebank had warned me, and even the locals say it has nine months of winter and three months of bad weather. But in July 1983 the sun shone without a break, the breeze was gentle, and the thermometer hovered around seventy degrees – which in Shetland is a heat wave. It was like Paradise with a temperate climate, silent and unpopulated and astonishingly beautiful. For three days after I settled into the hotel, I drove and walked – at first to see if the horror stories I had heard about the impact of the oil industry on the islands were true, then simply for the pleasure of it. I drove first to Esha Ness, which even the Shetlanders, who are spoiled in these matters, call a beauty spot. There is a lighthouse at its westerly point and behind that a huge dark zawn, its great cliffs reddish black and overhanging. On the shingle at the back of the zawn lay a crumpled black car and a bright pink fishing buoy, small as toys against the boulders. I walked around the rim of the zawn, then northward along the cliff tops past a headland tunnelled through by the sea to form a gloomy arch, past Drid Geo, Moo Stack, and the Villians of Ure, all of them pieces of rock of varying sizes that stick out from the sea just beyond the cliffs. The turf was smooth and springy, thick with sea pinks. Further north was another squatter, squarer arch, which looked as if the land had stepped forwards and planted a great booted leg in the Atlantic.

In two hours' walk I saw only one other person: a man marching purposefully towards me, camera and binoculars slung around his neck, rucksack on his back. We nodded to each other in a friendly way but did not speak, both of us unwilling to break the silence of this remote place. The only sounds were the waves beating on the rocks, the wind over the grass, and the gulls crying. When a dog barked, just once, from a croft inland, it sounded as loud and unexpected as a gunshot. At the Head of Stanshi, just beyond the Grind of the Navir – another lump of rock sticking out of the water – the angle of the cliffs eased and they became brownish slabs sloping peacefully down into the sea. Four seals were sunning themselves at the water's edge, but they saw me sooner than I saw them. One slithered immediately into the sea, arched, plunged, and disappeared. The others – a father, mother, and half-grown pup – slipped warily closer to the water, then lay back and went on sunbathing, watching me steadily all the while. It seemed a reasonable precaution considering the ominous little billboards outside cottages on the road from Lerwick advertising 'Sealskin Crafts'. The seals' coats were brownish, like the seaweedy rocks, and they lay with their tail flippers raised and neatly crossed, as though about to applaud.

The next day I drove south to Muckle Roe, a sparsely populated island that dangles below the handle of the Shetland dagger, connected to it by a brief causeway. The cliffs were lower and gentler than at Esha Ness and more vividly coloured: orange and yellow rock dropping into orange, sandy coves. The day was windless, hot and sticky; the sea glinted like sheet steel. I followed a vague footpath, up and down, across headlands, past little coves, while the nesting gulls screeched at me. A pair of black and white oystercatchers with long, thin, orange bills followed me for about a mile, crying unceasingly – a high, anxious note – circling, landing, circling, watching me intently all the time. When I returned to the cove not far from my parked car, there was a family group on the beach, enjoying the sun: two small, solemn children and three hefty young women, one with a thick braid of red hair down her back. A fourth woman had waded into the glassy sea and stood there hugging herself, the water up to the middle of her thighs. She was wearing what looked like white thermal underwear that left only her forearms exposed. She was perfectly still, outlined against the dazzle like one of Seurat's bathers. Suddenly, she ducked under the water – in and out, once only – and let out a great Scottish wail of anguish: 'Aaa – ooo.' Her companions watched in silence, waved to me when I waved, but uttered not a word. It was as though the emptiness and beauty of the place swallowed people up, making language an intrusion, an impertinence. In that engulfing nullity the social fabric would need to be strong.

The emptiness of Shetland seems uncanny at first until the reason for it becomes blindingly obvious: there are, quite simply, almost no trees on the islands, nor even any bushes. Remnants of dwarf birch and willow have been found buried deep in the peat, but the wind and the salt and the sheep destroyed them centuries ago, leaving a landscape without foreground, without focus, just rolling moorland dotted with sheep and broken occasionally by low, sculptured, curving terraces, like archaeological digs, where the peat has been dug out. Here and there brightly coloured sacks of peat wait by the side of the road to be collected.

Here and there, too, are little cemeteries, walled against the wind, walled as though for company. Each hamlet has its own but some have been set down seemingly at random in the middle of nowhere. The prettiest I saw was above a mere on the road to Esha Ness. At its centre was a big tombstone set upright in front of a grassy hump with walls of rough rock like a primitive family vault. A pair of gulls sat on top of the stone, peering around in a lordly, irritable way, as if they were the owners. Below them were carved two coats of arms and a great deal of small writing, most of it obliterated by the weather, ending sonorously, 'beloved in life, lamented in this death'. Below that was an indecipherable Latin tag and a row of eighteenth-century hieroglyphs: crossed bones, a coffin, an hourglass. A blue and white bus had been abandoned by the mere, its tyres gone, windows smashed, door rusted open. Near it, a group of shaggy Shetland ponies were drinking at the water's edge. They watched me uneasily, ready to take off, and one of them pawed fractiously at the water. Beyond the mere was a single farmhouse on a headland, then a wide bay with its folded coastline diminishing into the distance. If there had ever been a village in the area, it had vanished without trace.

Although the Shetland landscape is as smooth, undulating and featureless as a gently rolling sea, the islands geologically are the stump of a mountain range that was eroded millions of years ago. In outline on the map they have the same wild, jagged shape as the Alps seen from the air. There are bays and inlets by the hundred and of every shape and size, three thousand miles of coastline for five hundred and fifty square miles of land area. Because of this, and because of the sparse population, the roads on Shetland are unlike any other roads I have driven: they simply end. Even the main A970, which begins at Sumburgh on the southern tip and runs north for the whole length of the mainland, just fades away at its northernmost end. The name on the map where the red line ends is Isbister. But Isbister was three farmhouses, a walled graveyard, and a field of yellow irises. Beyond the farmyard the road continued unsurfaced, but there was a gate across it, and nailed to the gate was a notice that said, 'Private Road. No Dogs. No Entry.' I parked the car on the grass and trudged up the dusty private track, then struck off across the moor to the ordnance survey trig point at the top of Lanchestoo Hill. To the east, across Yell Sound, was the island of Yell, its low cliffs yellowy brown, the land beyond them flat and blank. Nobody – not even the man from the Shetland Tourist Organisation – can think of much to say in Yell's favour, except that it produces quantities of excellent peat. A big tanker was making its way slowly north from Sullom Voe terminal towards the open sea. The strip of unsurfaced road ran on northwards across the rolling moors to where the mainland of Shetland ends at a lighthouse, a single white tooth with a cluster of rocky little islets beyond it: Gruney, The Club, Fladda, Turla, Ramna Stacks, Outer Stack. When I looked back south I could see a scattering of houses on the edge of Burra Voe and the ribbon of the A970 winding off into the blue hills. But apart from the tanker, there was no sign at all of Europe's largest industrial development.

Even when I finally drove to Sullom Voe, there was no trace of the terminal until the last moment, when I turned a corner of the wide new road and saw, rising out of nowhere, a lone gas flare blazing into the sunlight above a low hill. Another corner and the road emerged on to the inner bay, Garth's Voe,

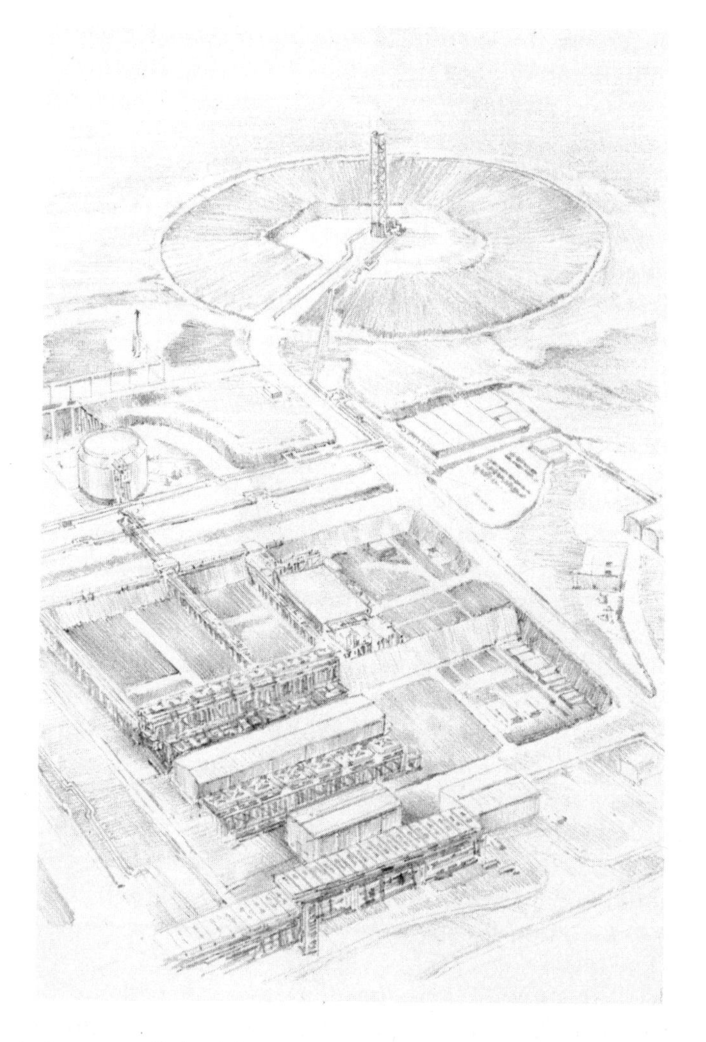

and there on Sella Ness, the headland that forms the Voe's southern edge, were the Port Authority's modest administration buildings, an airstrip, and a jetty where two muscular tugs were tied up. Across the bay was the terminal itself, a thousand acres of high technology, miles of pipes and roads, rows of storage tanks and buildings. Yet it hardly intruded on the gaunt landscape. Apart from the power station, the buildings were single-storey, and even the power station – the largest building in Shetland before the new sports complex at Lerwick was completed – was painted green, like many of the storage tanks, to blend into the moorland behind it. The whole vast enterprise seemed to have melted into the landscape. It was in place, it worked, yet it scarcely existed. Even Ian Clark could not have imagined that the view would have been so scrupulously protected.

North Sea oil arrives at Sullom Voe by two pipelines, one from Brent, the other from Chevron's Ninian field. Both are about a hundred miles long, both cost more than a million pounds a mile. The oil comes ashore as 'spiked', or 'live', crude – with the gases dissolved in it. The gas is then stripped out and separated into its five components. The two lightest fractions, ethane and methane, are used to drive the terminal's power station. (It can produce 125 megawatts, almost twice that of the power station at Lerwick that provides electricity for the whole archipelago.) The heavier gases, butane and propane, are cooled until they become liquid and are then shipped out in refrigerated tankers as LPG, liquefied petroleum gas. The fifth component, pentane, is spiked back into the crude for processing later. This 'dead' crude is then pumped into storage tanks where it waits to be shipped by tanker to the world's refineries. In other words, Sullom Voe is a holding area, a buffer between the North Sea oil fields and the refineries in the south. Its sixteen crude-oil storage tanks, each capable of holding six hundred thousand barrels, enable the offshore fields to go on producing even when the tankers are immobilised by storms.

Yet all this intense activity seems to occur without human intervention, as if by magic. The terminal was as unpeopled as the Shetland landscape, although the car park was full and the reception hall of the main office building made me feel as if I had been transported abruptly back to London: soft carpets and soft lighting, telephones ringing petulantly in the background, secretaries coming and going. But once outside in the terminal itself, there was no one at all, apart from the security guards on the main gates. For an hour I was driven around by Peter Guy, the ex-RAF officer who now lives on Yell, and we might as well have been in a lunar module guided by ghostly presences back on earth. Guy was cheerful and informal – the RAF is the least buttoned-up of the services – and full of facts. We went first to the terminal's jetties, a narrow gangway at the end of four long arms reaching out into the bay, one for the refrigerated tankers that take the LPG, the other three for oil tankers. That day there was only one giant tanker moored up, connected to the shore by the umbilical of a great loading arm. There was no one on its deck and no one on the jetty, although the oil, Guy assured me, was being pumped in at full blast – a maximum of twenty thousand tonnes of it an hour.

We drove back along a line of huge silver pipes into the terminal, where Guy ticked off the landmarks as we passed: the fire station, the power station, the de-butaniser, de-propaniser, and de-ethaniser, the electrical distribution substation, the process control room, the gas compressor house, the separator drums, dehydrators, heat exchangers, and the elaborate machinery in which the crude oil is stabilised – a thousand acres of tanks and cylinders and twisting pipes, steel ladders and girders, three and a half million metres of electric cable, low windowless buildings, pylons and steel towers painted silver. And still we saw no one.

It was different, Guy told me, when he first arrived at the time the terminal was being built. At the peak of the construction period there were seventy-two hundred people on the payroll and, even with more than a quarter of them alternated on leave, between four and five thousand were always working on the site in a jungle of equipment and with twenty canteens to keep them fed. Two prefabricated villages were put up to house this invasion, and when they proved inadequate the oil companies chartered two cruise ships that they moored in Garth's Voe. 'My first day here, I got lost in the chaos and nearly drove off the end of the jetty,' said Guy. 'And if I had, no one would have ever known. Weeks later perhaps, somebody would have said, "I wonder what happened to that chap Guy? Only stayed a day. Very odd." '

We drove up a hill at the back of the terminal and parked near the three steel fingers from which excess gas is flared off in emergencies. From above, the great maze of pipes and machinery looked like the intestines of a metal giant laid out for an anatomy class. Here and there, puffs of steam rose hissing into the air, and off to our right the main gas flare blazed in the sunlight with a sound like urgent, muffled drums. A helicopter, carrying core samples from an offshore platform to the terminal's laboratories, arced in from the north and dropped out of sight on to the landing strip at Scatsta. A single truck lumbered along a road on the far side of the terminal. Nothing else moved in the silence. Yet the whole vast plant was working flat out, controlled by technicians hidden away in the square, low computer centres. I understood now what Brian Lappin had meant when he said that oil is 'a capital-intensive industry that prides itself on its efficient use of manpower'. The terminal and its maintenance contractors have a total of six hundred and fifty employees on their books for day and night shifts, but not one of them was visible.

When I opened the car door to stretch my legs and look around, Peter Guy said uneasily, 'I wouldn't do that, if I were you. They're watching us on video in the control room at this moment.' I closed the door again quickly. Sure enough, on the legs of a pylon overlooking the road was a video camera that moved when we moved.

As we drove back towards reception, Guy began to talk about his family. His wife, he said, taught music in the local schools. They had a good life up here, peaceful and satisfying. Then he said, 'The Shetlanders believe in fairies, you know. They tell stories of trows – their version of trolls – who are sleeping in the hills, biding their time. We don't take it seriously, of course, not in this day and age. But you live here for a bit and it begins to make sense.' It made sense, too, in the eerie, silent emptiness of the terminal. With all that modern magic as the reality of his everyday life, he would need to believe in something.

There did not even seem to be much difference between the vast unpeopled terminal shut away in Sullom Voe and the trows hidden in the hills. Although the terminal has transformed the economy of the islands, it has altered their emptiness very little. It is as if the Shetlands had their own magic that swallows up people as it has swallowed up trees, leaving a green and temperate desert of grass and peat and rock, and the sea waiting wherever you turn. In the long summer days – at eleven p.m. it was still light enough to read without a lamp – the place was as serene as the Isles of the Blest. But in winter the Shetlands must be like the rock in Eliot's 'Dry Salvages' that 'in the sombre season/Or the sudden fury, is what it always was': a flattened mountain range rising out of the bitter northern sea.

Despite that, the islands cast a spell that is hard to break. 'When I left Shetland I vowed deliberately that I would only return on business,' said Ian Clark. 'I was so in love with the place and the community that I reckoned that if I didn't make a clean break I would never be able to settle anywhere else.' Clark had left because his success in dealing with the oil companies brought him offers he couldn't refuse. But there seemed to be many others who had originally come to Shetland solely because there was well-paid work to be had and were then taken over by the spirit of the place and could no longer get away. One morning on the road between Hillswick and Brae, I picked up a couple of hitchhikers, a lean, dark man in his middle thirties and a much younger woman, pretty but sulky. The man said he was from Southampton, although he spoke with a Yorkshire accent. He had come to Shetland eight years earlier to work as a joiner on the Sullom Voe construction site but had stayed on when the terminal was completed and now lived in one of a cluster of rundown caravans not far from the St Magnus Bay Hotel. It was three years, he said, since he had left the islands: 'I dunno. I just can't seem to get away.' He brooded for a while in silence. In the rear view mirror I could see his girlfriend watching us distrustfully. Then he brightened slightly: 'Of course, there's plenty of work and the money's good.' I wanted to ask why then, after eight years, was he carless and living in a lopsided caravan. Perhaps it had something to do with the sullen girl in the back seat who seemed to be younger than him by at least ten years; perhaps there was a wife somewhere and alimony payments. I glanced again into the rear view mirror; the girlfriend's eyes did not waver and her expression did not change.

'What's the public transport like?' I asked.

'Public transport!' he repeated scornfully. 'Two buses a week to Lerwick. No, three now. That's your public transport.' He lapsed once again into silence.

When I dropped them off at Brae the girl muttered, 'Thanks.' And that was all she said.

At the hotel they had told me that if I wanted to see what Shetland had been like before the oil came I should go to the western part of the mainland where the great, ragged pommel of the dagger juts out into the Atlantic. I drove south on the A970 to a hamlet called Voe, then took a meandering B road south and west across rolling countryside that was even bigger, emptier and more beautiful than the empty landscapes I had already seen. For mile after mile, the road wound up and down over bald moorland dotted with boulders, pitted with little lakes. Occasionally, the road dipped down to an inlet where a few houses, a jetty, and a couple of boats at anchor had a name on the map: Gonfirth, Souther Ho, East Burrafirth, Aith. Then the road was off again into the mild, empty hills where larks rose and fell, singing their hearts out, and sheep grazed peacefully. There are said to be three hundred thousand sheep on the Shetlands, but the crofters make do with flocks of fifty or sixty and have a name for each animal. Near Bixter the road flattened and straightened for a few miles west, then took off again into rolling country where every hollow seemed to hold a little tarn, its glinting surface ruffled by the wind, as if the sea had not yet fully receded. But without trees, without foreground, the landscape was vague, endless, hard to focus on.

Then suddenly the foreground appeared without warning in the shape of a battered old Datsun that arrived at the blind brow of a gentle hill at the same moment as I reached it from the opposite direction. The road was single-track, like most of the minor roads on Shetland, and the Datsun was squarely in the middle of it. I swerved on to the grass verge and came to rest with my car heeled over like a racing yacht, its nearside wheels firmly in a ditch. The Datsun skidded to a halt. Its coachwork was dented and patched with primer; there was a Conservative Party election sticker on its rear window. The driver got out unsteadily. He wore old jeans and no shirt. His plump body was pale and hairless, his mop of curls unkempt. I could smell the booze on his breath from two yards off. Yet the odd thing was, neither of us said a word. We simply stared at each other glumly while the larks sang overhead. Although years of city driving have conditioned me to shout first, then compromise grudgingly, for the first time there seemed to be no point in the charade drivers perform in order to protect their no-claims bonuses. He's drunk, I thought; shouting won't get me anywhere; anyway, I'm going to need him to get my car back on the road. But more important was the place itself: whatever happened to two pin figures in all that emptiness could not amount to much. So we looked at each other, more embarrassed than angry, and then we looked around. From where we stood, the road was visible for miles in both directions. Nothing moved on it. We were going to have to pull the car out of the ditch by ourselves. The shirtless man seemed unperturbed. On the evidence of the dents in his car, he had been here before.

It was like a film with no soundtrack. Silently, we went over to my car, got down in the ditch, and began to heave. Although his body looked flabby and unaired, he was very strong. The muscles knotted on his pale arms, his eyes bulged, his face ran with alcoholic sweat. We strained for ten minutes – first at one end, then at the other – and succeeded in manhandling my hired Ford six inches up the ditch. Then the sump wedged into the peaty turf and would not budge. 'Now what?' I asked. The man shrugged, walked back to his car, and began to fiddle with a CB radio on the dashboard. His voice was muttering and low, hard to catch. A gull swooped down on to the road and stood watching us disapprovingly. 'Nae luck,' said the pale man and shook his head. We went back to the ditch and heaved again, but with less conviction. I was wondering if I would ever see him again if he drove off for help, when the gull rose disdainfully and another car came slowly over the rise. The driver was elderly but heavily built, with fine white hair and a massive belly. Without saying a word, he climbed down into the ditch with us and began to heave. This time we got the rear wheels most of the way on to the verge, but still the sump would not shift. The elderly man stared at it, head cocked to one side, then took a length of rope from the boot of his car and pointed briskly at the Datsun. 'Back her up,' he said. 'We'll tie 'em axle to axle.' Within minutes, my car was back on the road. When I tried to thank the old man he cut me off with 'Nae trouble.' He and the Datsun driver muttered together in dialect for a moment, then drove away in convoy.

I began to understand what Ian Clark and the others had meant about the closeness of the Shetland community, and about how behaviour was affected by the exposure and loneliness of the place. In the emergency, my London conditioning had vanished. All that mattered was to get the car back on to the road; everything else was a waste of time and a waste of energy, a mere impertinence in that great empty landscape.

I followed the road to Sandness, then on to Melby, where it ended. Sandness was a scattering of houses and a sharp bend in the road. At the bend was an ancient petrol pump and a shop with a weather-beaten sign over the door: 'Robert Jamieson & Co. Licensed to Sell Tobacco'. At the corners of the sign, in old-fashioned script, were the words 'Grocers', 'Hirers', 'Drapers', 'Hosiers'. The shop was closed.

Melby was even smaller: two squat stone houses, a shed and a jetty. Four freshly painted clinker-built rowing boats were pulled up on the shingle – one white, one yellow, one grey, one a brilliant mauve. On the slipway next to the jetty was a rubber boat with a big outboard motor; beside it, a group of divers in black wet suits were sorting their gear. (The waters off Shetland are littered with wrecks and some of the divers I met later told me they came here in their time off to try their hand at treasure hunting.) Just offshore was an islet called the Holm of Melby, square and flat and featureless. Beyond that, to the north, lay the ragged coastline of Papa Stour, big and low on the horizon. But westwards, towards the misty sun, there was only the shifting grey Atlantic, and Greenland the first landfall. Apart from the divers with their rubber boat and flippers and bottled air, nothing seemed to have changed at Melby for a hundred years. And apart from a road and a shed and two stone cottages, the place was the same as it must have been when the Norsemen first arrived in the eighth century – beautiful, untouched, silent, stern. Greenland seemed closer to Melby than the terminal at Sullom Voe, and when I turned the car around and headed back east towards the evening, I felt I was going home, back to the twentieth century.

11

At the beginning of April 1977, Red Adair, the legendary Texas troubleshooter for the oil industry, issued a solemn warning about the North Sea oil fields:

> Whatever precautions are taken, there'll be a disaster in the North Sea, sooner or later. There are no proper facilities for coping with it. The thing is time, to get trained personnel there. By then the well may have caught fire – then it gets larger and larger, like a chain reaction, from this well, to the next well, to the next well. The more wells you have, on any platform, the more difficult it gets, because the heat will go on to the next tree. You get flames, leaks, and it just goes on and on. The hardware to deal with disaster? At the moment, for a real blowout, you don't have anything. Just a few little vessels, a thing that will squirt water. BP's Forties *Kiwi* is totally inadequate. I've been to England a number of times, and talked to them on the North Sea about the equipment they would need . . . Everything's sitting out there, wide open, with nothing to protect it.

Less than three weeks later, on 22 April, his predictions were fulfilled in the direst possible way in the Norwegian sector when Well 14 on Ekofisk Bravo blew out. The disaster occurred in the space of five minutes, between the removal of the

Christmas tree and the positioning of the blowout preventer. The preventer, a massive contraption weighing seven thousand pounds, was in fact in place, but it had been put in upside down. In the brief time while the operators tried to turn it right side up, the oil came gushing out of the wellhead on to the platform. It went on gushing for eight days at the rate of three thousand tons a day, until there was an oil slick covering nine hundred square miles of the North Sea. Had it caught fire, the platform might have burned for months. But Red Adair himself and two of his lieutenants – 'Boots' Hansen and Dick Hatteberg – finally recapped the well and brought the catastrophe under control with typically Texan aplomb. (When Hansen arrived at Stavanger in Norway he was asked if it would take thirty-five days to control the blowout. He answered: 'Thirty-five days? I only work thirty-five days a year.') The oil slick was dispersed by luck, broken up by the wind and waves, after four specialised Vikoma skimmers, specially developed to collect floating oil, had failed to do their job.

Adair had not only predicted the disaster, he had also submitted plans for a specialist firefighting vessel that the oil companies had rejected on the grounds of expense. But they changed their minds after the catastrophe, and since then a series of these vessels has appeared in the North Sea, the latest of them being MSV *Stadive*. When I first went offshore in March 1983, *Stadive* was about to begin her maiden voyage from the beach up the pipeline to Brent. She had been built in Finland and launched into the Gulf of Bothnia the previous November. The journey from Finland to Aberdeen had taken a week, with *Stadive* crawling across the North Sea in the dead of winter. But by the time I returned to Brent in July, she was already hard at work in the field.

Like *Treasure Finder*, MSV *Stadive* is a self-propelled, twin-hulled semisubmersible, a gigantic coffee table, three storeys high, supported by six columns and resting on two pontoons the size of submarines. MSV stands for Multifunctional Service Vessel, and one of her functions is as a Red Adair-style, mobile firefighting and rescue unit. She carries a 'wellkill' package and an elaborate access tower with a movable staircase for transferring men and gear to and from a stricken installation. She also has sixteen fire monitors of different sizes strategically placed around her sides. The monitors look like cannons, painted red, with heavy silver collars near their muzzles. But their bases are twisted like coiled snakes in order to distribute the battering force of the water. Between them, they can deliver ninety-two hundred cubic metres of water an hour, projecting it two hundred metres horizontally with a force that can smash through the wall of a module.

Like so many North Sea statistics, the figures are overwhelming and meaningless. They give no hint of the power and strange beauty of the monitors in action. A deep thunder starts up, as though somewhere far below your feet an underwater volcano had begun to erupt. The thunder grows louder, deeper, more intense, and the whole vessel vibrates like a creature in *tremor mortis*. Then it jerks abruptly backwards, as if punched, and suddenly grows wings: great ostrich feathers of water arcing far up into the sky, spreading out into a fan and gracefully falling. But because this is water that the cannons are firing, not shells, everything seems to happen in slow motion. After the backwards shudder of the vessel, you can actually see the silvery-white projectiles race skywards before the great wings of water unfold. It was at last a process commensurate with the forces of nature that the oil companies explore and harness with their technology. In modern oil production there are no more gushers, no more moments when it looks as if 'the earth had cut an artery'. The only indication of the scale of the operation is in those endless lists of figures with so many digits that they seem unrelated to ordinary life. But the sea cuts an artery when *Stadive* fires its monitors.

Firefighting is a function that Shell hopes will not often have to be performed by *Stadive*. In the meantime, she earns her keep as a diving vessel, moving around the field to wherever underwater maintenance and repair are needed. But firefighting and diving have one common requirement: the vessel must be able to stay where she is put whatever the weather is doing. To achieve this, *Stadive* has a Dynamic Positioning System

(DPS) made up of nine propulsion thrusters, six on the port pontoon, three on the starboard. As marine engines go, the thrusters are relatively small but very powerful: each of them puts out 2,400 b.h.p. (brake horsepower) to drive variable pitch propellers that, between them, can be turned through every angle on the horizontal plane. The coordinates of the vessel's required position are fed into a computer that operates the thrusters continually in minute bursts to counteract the effects of the sea. *Stadive* can come up to within a few yards of an installation and stay there safely in everything except the very worst weather, enabling her divers to work virtually all year round. They have even kept going in a force-ten gale, a feat they are proud of but would prefer not to repeat regularly.

'It couldn't be done without a computer,' said Alec MacLeod, *Stadive*'s captain (his official title is staff master). 'Two computer systems, in fact: one monitoring the DPS, the other monitoring the first.' MacLeod is a stocky, affable man with green-blue eyes and strong teeth. 'The last thing you'd want to do is take over manually,' he said. 'There is no possible way a man could keep the vessel in position accurately enough.' We were on the Navigation Bridge, which is also the Firefighting Control Centre. The enclosed base of one of *Stadive*'s three ponderous cranes filled the middle of the room; the outer walls were lined with shining banks of controls – inclined white panels covered with switches, buttons, knobs, dials and coloured lights. Through the forward window, Cormorant A loomed out of the busy water a couple of hundred yards off, a bright orange safety ship swinging close to her. The sky was a fragile blue. On such a beautiful summer's day it seemed hard to concentrate on all that technology, although without the technology there would be nobody in this remote place to admire the view. 'Don't let the sunshine fool you,' said MacLeod cheerfully. 'It's blowing a force four out there.'

Like the DP computers, *Stadive*'s control systems are also duplicated elsewhere on board. On the port side of the deck below the Navigation Bridge is the DP Room – more formidable banks of switches and dials and lights, and one man staring gloomily out of the windows at an empty sea. Opening off that is the far larger Operations Control Centre, windowless and even more jammed with elaborate, glimmering switchboards. Technicians moved purposefully about, lights flashed, the air hummed quietly. MacLeod showed me an instrument that looked like an outsized, intricately calibrated spirit level with three big dials. The needles on the dials were all slightly off centre, indicating that *Stadive* was listing a degree or two. He pressed three buttons, which promptly lit up. Equally promptly, the needles on the three dials began to swing – very, very slowly – back to their centre point. In a few moments MacLeod had shifted twenty tons of ballast in the pontoons and the vessel was once again riding dead level. The lights on the panel went out.

We took a lift down one of the six columns that support the superstructure, dropped smoothly below the waterline into one of the pontoons, and came out into a circular chamber so filled by the roar of engines that the air seemed almost solid. The thruster engines were about five feet across and hooded; under the metal hoods the parts whirred so fast that they seemed to stand still. On the side of each thruster was a device like the hand control of an old-fashioned lift: a semicircular steel plate from which protruded a little lever with a black knob on its top. The levers moved continually as the computers changed the pitch of the propeller blades in the water below. The chamber was as immaculate as an operating theatre and permeated by the sharp, rousing smell of clean machine oil.

For most of the time, all this complex, state-of-the-art technology and banks of computers monitoring other computers are there to serve just one purpose and two people. Suspended from the moonpool at *Stadive*'s centre far down in the black silence of the seabed, was a manned diving bell; attached to that by an umbilical, a diver was working on the pipeline from Brent to Sullom Voe, all his movements monitored on the instruments and video screens in the small, crowded Diving Control Room immediately behind the Operations Control Centre.

Maintenance of one kind or another is a full-time job in the North Sea. Dragging anchors from fishing boats damage the concrete casing of the pipeline, or the seabed shifts in the currents, leaving unsupported spans of pipe. And the installations themselves, constantly deteriorating in the wind and the rain and the salt water, must be inspected at least once a year. To do this, the seaweed, barnacles, and rust must be blasted off with high-pressure hoses until the welds are cleaned to what is called 'bright metal', when they can be tested with sophisticated ultrasonic and magnetic particle inspection tools. It sounds like the North Sea mixture as before – dazzling high tech plus a great deal of elbow grease – until you remember that most of this work has to be carried out far below the stormy surface of the sea.

The maximum depth at which divers are legally permitted to work while breathing ordinary air pumped from the surface is fifty metres. Because air is eighty per cent nitrogen, and nitrogen under pressure is a depressant, like alcohol, nitrogen narcosis sets in below fifty metres – and often sooner. The mind wanders, reactions slow, vision blurs, simple jobs become intolerably difficult, simple jokes provoke hysterical laughter. Air diving has a further disadvantage: when the diver returns to the surface, decompression is difficult and slow, the danger of 'the bends' is always present. (When a diver goes under pressure, the gas that is pumped into his lungs then finds its way into his bloodstream. This happens quickly when pressure is applied, but the reverse process – 'coming up for air' – by which the lungs filter the gas out of the bloodstream, is much slower. When the pressure reduces too rapidly, bubbles of gas form in the bloodstream, get trapped in the joints, and cut off the flow of blood. 'Agonising,' a diver told me. 'It feels like sharp needles in your ankles, knees, elbows, and especially in your back.') The only way to avoid the bends is by extremely slow and careful decompression, which makes air diving a costly process. 'A man would go in, make his dive, and maybe work on the bottom for an hour. Then it would take days to decompress him,' said John Roberts, diving officer for BP, whom I talked to in Aberdeen. Roberts is a big, beefy man with a slightly drooping face and a wry, self-knocking manner that make him look like a sturdy British version of Walter Matthau. He learned his skills in the Navy and was diving until he was forty-six – making him the oldest diver in the North Sea. He stopped only because he injured his leg in a road accident. He now walks with a slight limp. He went on: 'In the old days – in diving terms that means ten years ago or less – we used what is called standard gear: a copper helmet, big lead boots, big lead weights front and back, and as many clothes as you could get on under your canvas suit. We used to go down and crawl around on the bottom on our hands and knees most of the time. Then the oil companies came along. They wanted people to stay down for longer periods, and they wanted greater depth. So suddenly a great deal of money was poured into the business.'

Out of that investment came hundreds of new developments and discoveries, until diving depths that had once set records and cost lives to reach became considered everyday working levels. One of the most significant discoveries was that when divers worked at depth for extended periods, until their blood became, literally, saturated with inert gas, their decompression time was not significantly longer than that of divers working for much shorter spells. It was therefore logical, as well as economical, to create an environment in which divers could live at the atmospheric pressure at which they were working and decompress only when they were ready to return to the beach. As always in the oil industry, the step from economic logic to practice was short and decisive. Divers now spend roughly four weeks in saturation, living in a specially pressurised habitat on *Stadive*'s lower deck, and moving directly from it to the similarly pressurised diving bell that takes them down to the seabed. They work for twenty-one days in eight- to twelve-hour shifts, without ever breathing ordinary air, then decompress for three or four days. At the end of that time, they spend four well-earned weeks back on the beach.

The problem of nitrogen narcosis, *la folie des profondeurs*, has been solved by using a mixture of helium and oxygen instead of air. Helium is a totally inert gas that does not misbehave under pressure and is easily absorbed into the bloodstream and body tissues. But this makes it hard to get out and the process of decompression comparatively slow. It also has the great disadvantage of high thermal conductivity, taking heat away from the body six times faster than ordinary air. Diving suits have been redesigned accordingly. Gone are the hammered copper helmet, lead boots, and canvas suit in which John Roberts learned to dive in the Navy. In their place is a stylish, double-skinned outfit in black and red that looks like an overweight ski suit. Threaded all through it, and extending into the gloves and feet, is a dense network of pipes, like an anatomical map of the blood system, through which hot water circulates. The water is, in fact, very hot; it reaches the diver at 50 to 60 degrees Centigrade – between 120 and 140 degrees Fahrenheit – although, because of the helium, it feels to him only pleasantly warm. It is pumped through the umbilical that connects him to the diving bell and comes out through tiny perforations in the pipes, flushing the layer between the inner and outer skin of the suit, and exiting through a hose at the wrist. Under the suit the diver wears only a lightweight thermal outfit – called a 'woolly bear' – to protect his skin from abrasion. His helmet is streamlined plastic; he wears flippers on his feet.

The umbilical is a thick, rubber-covered rope made up of five strands: the hot-water line, a strain member that keeps the rope from breaking, a gas line that pumps oxyhelium into the diver's helmet, a pneumohose that measures depth, and a communications line connecting the diver's microphone with the Diving Control Room on the surface. But it is also an umbilical in the true sense of the word: without the hot water the diver would be dead in minutes. 'It's the cold that kills them, not the pressure,' Roberts had said. 'Hypothermia.' Strapped to the diver's back is an emergency gas supply and gas heater, called the 'bailout', that he switches on if the gas supply fails or the umbilical springs a leak. 'We've got about six minutes of air in the bailout,' said another diver. 'This is reduced a bit because of the size of the helmet, but it's enough to get back to the bell. If something goes wrong, the adrenalin starts pumping and you'll get back, believe me, in six minutes. Or a lot less. You'll breathe anything down there to get back.'

What it must feel like to be under five hundred feet of water, alone in all that darkness, is not something an outsider can easily comprehend. But what it looks like is plainly on view on two video screens surrounded by dials and switches in the Diving Control Room. In comparison with the constant coming and going in the DCR, the chat and the orders, the amplified rasp of the diver's breathing and his squawking voice over the intercom – he sounds like Donald Duck, another side effect of helium – life on the seabed seems curiously calm and slow. On one video screen is the working diver, on the other is the bellman, although at first glance all that can be seen of him is a pair of hands holding a fat paperback and surrounded by what looks like a mass of seaweed outlined against the sun. This eventually resolves itself into the bellman's legs, a complicated tangle of pipes and gear, and the open bottom of the bell with the lights reflecting off the water.

The diver, when I first saw him, had just finished sandbagging an unsupported span of pipeline and was trying to fix in place a job-identification board that he would then photograph for the records. But each time he propped the little board on top of the pipe, some obscure underwater current toppled it off again. After three tries, he managed to get the board steady, then backed away and pointed his bulky camera. A fish swam into view, flicked its tail, and disappeared into the murk. The board toppled over. 'Faaa . . .' said the diver.

'Don't let it getcher down,' said the DCR supervisor, who wore gold chains around his neck and one gold earring. His hair was curly and carefully styled, he had a Sitwellian, parrot's nose, and a Liverpool accent as thick as Ringo Starr's. He adjusted a dial while he talked, and watched the figures at the bottom of the diver's screen. There were two lines of them and those on the top were constantly changing: 'Depth 0143.3 (.4, .3, .4); Pitch – 19 (varying from −25 to +5); Head 000 (steady).' On the bottom line were the date, time, and job number: '07–05-83; 09:08.56; Inc No 883.'

The diver propped up the board again and backed off with his camera. The control room was silent except for the steady, distorted rasp of his breath over the loudspeaker; 'Huuuaaa . . . huuuaaa.' Then the sound of the breathing suddenly stopped. The supervisor grabbed the microphone and made frantic signs to his assistant. At the same moment, the breathing started again.

'Don't *do* that to me,' the supervisor shouted.

'Do whaaa . . .?'

'He must have been holding his breath to take the shot,' said the assistant.

The diver collected the board and began to swim slowly forwards up the pipeline, his trailing umbilical stirring up clouds of silt in his wake. A large spider crab was plastered against the base of the pipe. 'I fancy that for my dinner,' said the supervisor.

'They're right bastards to catch,' said his assistant. 'You get 'em behind the head and they still nip you.'

The supervisor shook his head wearily. 'It's a gang of bloody cowards I've got here,' he said.

The diver clambered up on to the top of the pipe and balanced there, a little unsteadily, profiled against the lights from the bell in the distance, a blue-black cutout with a halo of brilliance, as strange and lonely and vividly lit as a dream. The busy, enclosed Diving Control Room, with its dials and switches and obscure electronics, seemed a curious setting for beauty, and for a moment the whole enterprise changed focus. The men and machinery on the surface, the oil steadily pumping from miles down under the sea on its complex way to the filling stations seemed beside the point. The point was this one man going imperturbably about his business with four hundred and seventy feet of black water above him. Then he stepped back down from the pipe and was swallowed up in a cloud of silt.

We could watch him from *Stadive*, despite the engulfing darkness, because he is followed everywhere by a remotely operated vehicle, an ROV, sometimes called 'an eyeball', more generally known as 'Snoopy'. The ROV is an orange plastic ball, a couple of feet in diameter, with lights and a television camera on one side, a propeller on the other. It has an umbilical that connects it to an operator on the surface, who jiggles a little joystick to move it in any direction he wants. ROVs are

used instead of divers for the preliminary, routine inspection of underwater structures, moving, pausing, photographing, as the experts above require. 'An ROV saves a lot of money,' Roberts had said. 'You don't have to decompress it for days on end, it never goes sick, it never asks for a pay rise.' But the instrument's most important function is to monitor the diver, lighting up the place where he works and keeping a watchful eye on his wellbeing. When the ROV was first introduced, the divers resented it, feeling they were being spied on. 'If something goes wrong – say, you drop a tool – everyone can see. That's why they called it Snoopy,' said a young diver named David Hamm. But with use, the name has become affectionate and the divers now think of the ROV more as

Charlie Brown's dog than as Big Brother. 'It's cold and dark and lonely down there,' said Hamm. 'Snoopy's like a pet. You take it with you wherever you go. It's company, it's got lights, you can swim for it. Even though you know it can't help, it's still a comfort. It means there is someone up there keeping an eye on you. It makes you feel safe.'

Hamm was twenty-three years old – the youngest man I met offshore – short and trim, with close-cropped hair, a rectangular face and large, studious spectacles. Despite his youth, he already had five years' experience in the North Sea. He began diving seriously as a sport when he was fourteen; at eighteen he took a four-month air training course at a diving school in Los Angeles, then went to Fort William, in the Scottish Highlands, for six weeks of bell training, before going offshore. 'The funny thing is, I couldn't swim properly until I was sixteen,' he said. 'I could do a couple of lengths – you have to in order to qualify for the course – but even now I'm not much of a swimmer. Yet I could dive – swim with fins – since I was six. It still feels all wrong without fins.'

When I asked John Roberts what really deep water did to the diver, his face assumed a particularly Matthau-esque expression and all he said was, 'Not a lot. The deeper you go, the slower you move. That's all.' David Hamm, being many years closer to whatever it was that had first drawn him to this strange profession, was more forthcoming: 'Commercial diving is quite different from what I'd imagined. I'd always wanted to go into saturation and when I did my first deep dive the sensation was worth the investment. It's very eerie, like stepping onto the moon. I suppose that's what it is, really – an unknown world, but downwards instead of up. I have to admit that, when I started, I found the transition from air to deep diving very . . . stressful.'

'Stressful?'

Hamm blushed. 'Okay, frightening. The water is pitch black and you get a tremendous feeling of agoraphobia. Claustrophobia doesn't worry me at all, but when you get out of the bell in mid-water with all that water above you and below, and you can't even see the job . . . Well, whatever they say, everyone has a fear of the swim. You go out of the bottom of the bell, climb up on to its top, and leap. That climb only gains you a couple of feet, but it *feels* more. And it's like leaping off into the void, into nothingness. That's when I think about the depth of the water, although some of the lads aren't bothered at all. Okay, you've got a tether, the umbilical. But it's black as midnight and kind of directionless. After you leave the bell, you're on your own and it's a matter of pedal power. With all the diving gear – your helmet, your waterlogged suit, your bailout cylinder, and the umbilical dragging after you – a long swim becomes very strenuous. By a long swim, I mean more than twenty metres. In fact, the umbilicals are forty-four metres, although the new ones are sixty – and that's one hell of a long way. The other difference between deep diving and air diving is that down there everything is so huge. The legs of the cement structures stand on cement storage cells, and the domes of those cells are enormous; it's like swimming over the Pennines. It's the same with the steel structures; their latticework is gigantic and terribly confusing.'

In the circumstances, 'stressful' seemed a mild word for the activity. Yet when I asked him how long he would go on diving he shrugged and said, 'Till I'm thirty, maybe.' At the age of twenty-three, perhaps, thirty seems as far away as eighty. 'You see, you get so psyched up when you dive,' he continued, 'and it becomes such a big part of your life that I couldn't imagine cutting it off, just like that, and settling down to run a pub. Maybe I'll go into the management side of diving. Or maybe I'll just keep on diving.'

We talked about diving while he took me on a tour of the saturation chambers where the divers live when they are at work. *Stavide* has the largest commercial diving spread in the world: two bells, a small submarine, a divers' escape lifeboat, and room for twenty-eight men in saturation – two six-man chambers, two eight-man chambers. Hamm took me first to the surface decompression chamber, where an air diver goes if he gets the bends. It was a small cylinder, about fifteen feet long and five feet in diameter, like a punishment cell, with enough room to lie down but not to stand up. Along one side was a grey metal bench, with a telephone at one end of it, a bib (an oxygen line and mask) at the other. Opposite the bench, lashed back neatly against the wall, was a narrow bunk.

'It's just as well claustrophobia doesn't go with the job,' I said.

'The way things are nowadays,' said Hamm, 'the bends don't, either.'

In comparison, the saturation chambers seemed spacious and comfortable. The working divers are brought up to the surface in the bell, which locks on to a wet chamber, where they take off their suits. They then crawl through a hatch into the crowded area that passes for their bathroom: a wash basin, a shower, and a toilet. But in saturation nothing is simple, not even washing. There is a difference in pressure between the chambers and the outside, so no taps can be turned on without first phoning control to make sure that the water is correctly pressurised. To keep the air sweet, there are aluminium canisters next to the toilet containing fans that scrub the atmosphere of carbon dioxide: a filter of soda sorb granules lies above the fans; above that is a second filter, of carbon granules, to absorb the smells. And to cope with the humidity that builds up quickly in such an enclosed space, there is a kinergetic system that heats the air, then cools it, thus condensing the humidity and precipitating it into bilges from which it can be pumped away. Yet even with all these precautions, things still go wrong. Hamm told me he once did a spell in saturation when the sanitary area was unheated. 'A helium environment is the bitterest cold you can imagine,' he said. 'We did our best not to wash or use the loo the whole four weeks we were in sat. We weren't too fetching when we came out.'

The living quarters seem like the inside of a giant cannon: a long, tubular chamber with four bunks, in pairs, on each side of the curved walls – two up, two down – an oxygen bib above each. The colours of the walls and curtains are soothing blues and greys. The dividers between the bunks and the ventilators are aluminium. To protect against fire, the lighting is controlled from outside, the lamps are shielded, there are no switches. There is also a good deal of firefighting equipment, as well as water sprays in the pipes that run along the ceiling. Fire is the abiding fear, for in an oxyhelium atmosphere chemical sprays are of no use. 'The only alternative to water is ox blood,' said Hamm. 'But that's not as gruesome as it sounds. Under pressure it becomes just another white foam. It's very effective but it stinks something awful.'

Smoking, of course, is forbidden. 'But the odd thing is, even heavy smokers don't miss it when they're in sat.,' said Hamm. 'Maybe it's because you're being filled full of helium, or maybe you make a physiological adjustment as you come through that hatch at the beginning of a stint. But as you decompress, the shallower you get, the stronger the craving becomes. All it takes is for someone to say the word tobacco and immediately you want to smoke. And when the hatch opens again, the first thing you do is light up. We all say we're going to give it up one day – but only when they nail the lid on the box!'

Although *Stadive*'s saturation chambers can accommodate twenty-eight men, the normal operating number is twelve: three three-man teams – two men in the water, one in the bell – working around the clock in eight-hour shifts, and one team decompressing. With just Hamm and myself in it, the chamber seemed large enough. But with eight weary men?

'It's okay when you're in working mode,' said Hamm. 'After eight hours in the water, all that's left is exhaustion. Everything goes into slow motion. Having a shower is a major event; it takes ages. And sometimes you sleep for twelve straight hours. After a bell run, they give you an hour to clean up, then they feed you. The meals come in through an air lock on the side and you're encouraged to eat as much as you like – particularly good, nourishing things – because you use up so much energy in the water. We keep tea bags, Nescafé and condiments in here. If you wake up and want a cuppa while the others are still asleep, they send you in hot water and a big sticky bun.' His eyes behind his studious glasses became as wistful as a schoolboy's dreaming of the tuckshop. 'Those sticky buns are really delicious,' he said.

Even so, I said, four weeks is a long haul.

He shook his head. 'When you're working time moves really

fast, perhaps because you're sleeping so much. Decompression is much worse, because there's nothing to do but sit and wait. You read a lot or maybe you play cards. But then your companions' little mannerisms can begin to get on your nerves. Or a rattle somewhere that won't stop. Things like that. You have to learn to control yourself or life would become intolerable. But of course, you're getting psyched up to go home, so it's not so bad.'

Hamm clambered agilely out of the chamber and I struggled after him: head first, grab the top of the hatch, swing your legs through. Out to the noise of engines, a metallic voice over the intercom, a technician whistling as he adjusted an air lock on one of the empty chambers. Even the faintly oily air smelled good. And I had been in the chamber no more than half an hour. 'I'll tell you something,' Hamm said with feeling. 'Climbing out of that hatch is the nicest sensation in the world.'

Later, I asked another diver about living in saturation. Shane Beary is a Rhodesian who joined his country's army, then learned to dive in the South African Navy. He has worked in the North Sea for seven years and commutes each month from Brent to Marbella, in southern Spain. ('That's not much more difficult or expensive than getting to some places in England,' he said. 'And in winter it's dead cheap and much nicer.') He is in his middle thirties, a big man with thick blond hair, a blond moustache, and an opulent Costa del Sol suntan, but tough under his sleek exterior and under no illusions at all about the glamour of his profession. Neither does he underestimate its strengths. 'We're a very tightly knit group,' he said. 'I've worked with nearly all these guys, so there's a kind of link between us, a closeness. Maybe there are some people I'd prefer not to be with in sat. It's a question of personalities. But you learn to control your likes and dislikes, and not let little things get on your nerves. A very high proportion of us are ex-military. This means we have greater self-discipline than average. We can give a knock and take a knock; we know how to live together under pressure.'

But there are pressures and pressures, I said; back on the beach none of us could tolerate being shut up in a confined space for four weeks, even with someone we loved – particularly with someone we loved.

'Of course not,' he answered. 'If you were told to stay in a room as long as you could, but you knew that all you had to do to get out was open the door, how long would you last? In diving you know there is no way you can break or bend the rules, no way you can get out. It doesn't matter what happens back on the beach – your child could die, your whole family could be wiped out in an accident – it would still take you four or five days to decompress. So you make an adjustment.'

Beary might also have been describing the atmosphere on *Stadive* itself: good-natured, forbearing, preoccupied; a tightly knit group that has made an adjustment. The crew is smaller than that of most installations – there are berths for only 128 – and far friendlier. They greet you in the corridors and are helpful in a casual way. But they keep it casual because there is a continual turnover in faces as the specialists are choppered in and out to perform their arcane functions: computer men, radar men, electronics men. Occasionally, someone would appear in what looked, in the North Sea, like fancy dress – wearing a grey flannel business suit and carrying a black leather attaché case. And for a couple of days there were two Frenchmen on board, technicians from Comex, the Marseilles-based diving company; both wore stylish moustaches, and both sat for hours in the mess hall, doing sums on their pocket calculators and picking disdainfully at the lavish food.

Yet despite all the coming and going, *Stadive* felt more like a ship than an installation. Indeed, this great, awkward, oblong structure was often mysteriously on the move. There was no sense of forward motion, but sometimes the tone of the DP thrusters changed and became more purposeful, and up on deck a line dangling overboard trailed gently back at an angle, indicating movement. Each morning a different platform appeared off our beam as *Stadive* made its slow way along the pipeline: North Cormorant, South Cormorant, Brent Bravo with *Treasure Finder* low on the water alongside it, linked by the widow-maker, like a baby elephant holding its mother's tail. And always off on the horizon was the vague shape of an installation swaddled in mist, with its plume of fire rising, seemingly detached, into the clear upper sky.

'Diving is simply a method of getting a chap to his work site,' John Roberts had said. 'And air diving is a much nastier way of getting to work than deep diving. Think about it: it's the middle of the night, it's snowing or raining, blowing a gale, and damn cold. But you've got to disregard all that and get on with your job. In those conditions, a four-hour shift in the water is a long, long haul. Believe me, there's nothing glamorous about it. Okay, I suppose it's dangerous. So what? To me, that's like saying coal mining is glamorous. It's just uncomfortable and miserable and damn hard work. Deep diving, at least you're out of the weather.'

If diving is just another way of getting to work, then the diving bell is an elevator that sits in the moonpool above the open water at the centre of *Stadive*. The bell is a circular chamber, with a flat bottom and a domed top, inside a cursor – a skeleton of thick white tubes that come together at the top in a kind of open-weave cap. The bell's umbilical goes through the cursor, and the cursor runs on tramlines to eighteen metres below the surface, enabling the bell to be lowered smoothly through the disturbance of the upper sea and drawn smoothly back up again into the belly of the vessel.

A diver called Billy Moran explained the procedure to me. Moran, who lives in Somerset, was straight-backed, short-haired, and looked fit enough to go on diving another sixteen years, until he reached fifty, the obligatory retirement age. 'We leave the main saturation chambers,' he said, 'climb into the bell, lock it off, and trolley ourselves into a position in the centre of the moonpool where they can lower the bell into the water. As they lower us down, we read out the depth to them every ten metres. When we reach working level we'll probably hear a hiss of water. That means the pressure outside is the same as the pressure inside, and the seal on the bottom door has automatically been released. There are two doors in the base of the bell with a gap between them. We equalise the pressure between the two, open the top door, lash it back, open the bottom door, and let it fall away. We've then got access straight through to the water. The next stage is to get the divers dressed and do our checks. We're already wearing our suits, so all the divers need to do is put on their helmets and bailouts and sort out the working gear they're going to need. Then they're ready to go out on the job.'

The inside of the bell is high enough to stand in comfortably, but the equipment jammed into it makes stretching a problem: umbilicals coiled like giant sleeping pythons, helmets, bailout packs, a canister for heating, another canister containing emergency gas. The walls are a maze of pipes and gauges and dials and levers. Facing the gauges is a small button seat for the bellman. I said that on the videos the bellman had seemed to do nothing except read. 'Quite right,' Moran answered. 'He's only needed if something goes wrong. Then he has to get out into the water and try to pull the diver back to the bell with the umbilical. He has a lightweight plastic helmet for that, not a big hard hat, so his head will get wet and he'll probably feel a bit cold; but he's only out there for a few minutes. The rest of the time he just sits there and keeps an eye on the gauges. It's as boring as hell, so you need a good book. The truth is, you develop a feeling for trouble and it becomes second nature after a while: everything sounds right, so everything's going right. But you have to do checks every ten minutes. You sit reading a book, then take a quick glance around. You know where all the needles should be on the dials. For instance, you can watch the diver's breathing on the helium reclaim gauges: you set them for eight to ten bar and the gauge flickers about two bar down every time he breathes in, then flicks back again. If something goes wrong, you know immediately. Also, we've got repeats on the surface; if we miss something, they won't. The divers out here have complete faith in the people looking after them, from the man in the bell all the way up the line to the guy running the dive at the top. That's what it's all about, really – like one big family. There's discipline in the job, but it's all on a friendly basis. We trust each other; if we didn't, we couldn't do what has to be done.'

Family feeling, however, doesn't extend to the bell itself. 'A particularly cumbersome affair,' Shane Beary called it. 'Five years out of date. Technology is evolving at a huge rate, but they don't keep up because it costs too much and they're reluctant to spend money on things that don't show a direct profit. In the seven years I've been working out here, the diving systems have been changing all the time, but not once for the divers' comfort. On the contrary, the more they change, the more cluttered the inside of the bell becomes. What they're doing is filling a round bell with square equipment. You get three men coming up after a day's work, all of them tired and strained, and they have to go through this involved routine. There are all kinds of valves that have to be turned in sequence. That sort of thing. None of it is difficult in itself, but none of it is easy in the circumstances. Let's just say that it's unnecessarily complicated – particularly when you think of the clever things engineers can do.'

The things they can do underwater extend now to most of the things they can do on the surface, from simple labouring jobs, like packing unsupported spans of pipeline with sacks of grout that will solidify into cement, to sophisticated technical tasks, like taking profile measurements of a platform leg or using a strobe camera to take stereo photographs or manipulating an elaborate machine that bevels the ends of the pipe. They also have a mobile submarine chamber – demurely called a 'habitat' – for underwater welding. When not in use the habitat sits on the upper deck of *Stadive*, a ponderous steel box, painted white, with a big protruding steel beam across its top. From the beam hang four gigantic lobster claws, two on each side. The inside of the habitat is a square steel chamber with blank walls broken only by five sealed magazines for equipment and two large portholes – one on the top, one near the floor on the side. There is a round hatch near the lower porthole and doors at either end, opening on to the lobster claws. If an underwater pipe breaks, the habitat, its floor and doors retracted, is lowered over the fracture on guide lines, like an elevator in a shaft. Hydraulic jacks level the structure on the seabed, the floor closes up, the lobster claws grip the pipe and align it, the end doors close around it, and the water is sucked out of the habitat. A diving bell then locks on to the top porthole and the divers who are to do the welding climb inside the now dry and watertight habitat. All their delicate welding gear is lowered down to them separately in an airtight module that locks on to the porthole near the floor. The module is known as MISS – *Module d'intervention soudure sous-marine* – and it looks like one of the leggy, Heath Robinson contraptions that were landed on the moon: four orange tubular legs supporting a white, steel, oblong box with canisters, like shoulders, on its sides, and another larger canister, with bulges on it, hanging between the legs. It is a cumbersome piece of equipment that looks as if it had not so much been designed as evolved in use, dive after dive – a new wrinkle for each new need.

I spent about an hour in the habitat, up there on the sunny deck, although there was nothing much to see inside it. Perhaps *because* there was nothing much to see inside it, except the seven closed hatches of varying sizes that punctuated its metal walls. It was like a cell somewhere deep under Orwell's Ministry of Love. When I pulled the entry hatch to, the darkness was absolute and I began vaguely to understand what the blackness and claustrophobia of the seabed must be like. It was like being buried alive. 'Pitch black and kind of directionless,' David Hamm had said. He had also said with, in the circumstances, a curious formality, 'It is very hard to comprehend the size of those things down there.'

What makes people dive? I asked the question several times but none of the replies was particularly satisfactory.

John Roberts had shrugged and answered in one word: 'Greed.'

'The money comes in handy,' said Billy Moran. 'But I enjoy the work anyway. Even if the pay wasn't so good, I'd still do it. I couldn't stand all day in a factory or sit on my backside in an office. Okay, the first time I did a deep dive I found it a bit nerve-racking. I suppose I was a little awestruck, thinking, This is me down here. I hadn't really believed I'd ever do it, and there I was. Now I just accept it. It's my job and that's all there is to it.'

'Most divers have had the same lifestyle since they were

sixteen,' said Shane Beary. 'With that background, there's bugger all else we can do. It's an exclusive club, and once you join it there aren't too many alternatives for you.'

The clearest answer came from Ric Wharton, who has dived for fun but never for a living, and now runs an Aberdeen-based company called 2W Diving. Wharton is in his forties, thin-faced, bearded, restless, and slightly sporty in his appearance. His delicately checked suit and striped silk shirt, open at the neck, made him look more like a gambler than a millionaire businessman. But that was appropriate since in 1981 he had masterminded the biggest and most improbable diving coup of the century: the salvage of 431 bars of gold – each worth a hundred thousand pounds – from the torpedoed wreck of HMS *Edinburgh*, eight hundred feet under the Barents Sea, far above the Arctic Circle.

Wharton began to dive in the Scuba Club of the Imperial College of Science, in London, where he trained to be an engineer. 'In those days,' he said, 'you made your own wet suit, learned to dive in flooded gravel pits, and froze to death. An altogether miserable way of spending your time, but risky and interesting.' By the time he graduated, diving was an addiction, civil engineering merely a job. So he took work where he could be near the sea. First, he designed and built a lifeboat station on the north coast of Cornwall, then, in 1970, he joined an American company as a diving engineer: 'I spent three and a half months offshore in the Persian Gulf, then got ten days' home leave, all for six hundred dollars a month. In those days you got *used*. One morning I woke up and thought, This is no good; I've got a young wife back home and I'm selling my life. So after a year I joined Comex – by some accident of childhood history I happen to speak fluent French – and became managing director of their British company in Great Yarmouth. We built that up from one hundred and fifty thousand pounds annual turnover in 1971 to a twenty-five-million-pound turnover by 1976. But I'm an ambitious man and, like any entrepreneur, I got fed up with making money for other people. So in 1977 Malcolm Williams and I founded 2W Diving. When we started it was just the two of us, with one secretary. Now we've got one of the four biggest commercial diving companies in the world, with about one hundred and fifty people in our Aberdeen HQ and about a thousand on our offshore payroll.'

Wharton is now an international tycoon, who spends more time in aeroplanes than in the castle near Aberdeen that he has renovated for his family. But the passion that brought him into the business in the first place has not diminished. 'It is exciting, always exciting,' he said, 'because you are dealing with real things and taking large risks. You're also dealing with the sea, which is continually fascinating and mostly very frightening. A great untamed area, lovely one day, deadly the next. And because it acts in ways that not even the experts can predict, we are always breaking new ground, doing things that no one has tried before. We are repairing structures and developing techniques, as we go along, for complex operations that would be difficult on the surface, let alone five hundred feet down. Diving is fascinating in the same way. It's an adventurous profession that takes you to places where other people don't often go. There's also the pleasure of the company, of the companionship. Divers are a wonderful crowd of people. They used to be a very wild and ragged bunch, and to some extent they still are. But they are also exceptionally highly trained and professional. They have to be technicians, with a detailed understanding of the specialised underwater equipment, and they have to be skilled mechanics to use it. They also have to be strong as oxen to do the labouring jobs. Above all, they must be self-sufficient because, however technical and sophisticated the work, it's dark and filthy down there and they are on their own at the end of a hose. When a job has to be done on the surface there is always a certain amount of teamwork involved; if something goes wrong, you can get help from other people. The seabed is different. A diving vessel with its crew and its fuel and its diving team is going to cost twenty to twenty-five thousand pounds a day – over one thousand pounds an hour just to keep it there to do a job of work. You've got all that fantastic technology, but hanging below the ship is a diving bell, and hanging out of the diving bell at the end of an umbilical is a diver. And at the end of the day, everything depends on his know-how and his hand-eye coordination. Either he performs or he doesn't, and his lack of performance is exceedingly expensive. So if he's no good, he doesn't survive. The same applies to the team running the dive. For every guy in the water there are a lot of other guys up top in support, and the diver's life is totally dependent upon them. If they make one wrong move, they could kill him – him and maybe another dozen men in saturation. The possibility of a catastrophic accident is always present. It is a testimony to their professionalism that, despite the intricate technical chain, accidents hardly ever occur. The general public thinks of divers as being wild, hairy, heavy-drinking guys. And it's true that a fair bit of their income *is* consumed by wine, women, and song. (Why else have an income?) But when they are working they are deadly serious, utterly professional.'

Part of that professionalism, however, is in maintaining a casual tone and ignoring the utter improbability of the situation. One afternoon in the Diving Control Room I watched a diver inspecting a damaged section of pipeline between Brent Delta and Brent Charlie. The scene was brilliantly lit, as though in a studio setup, while Snoopy's camera panned slowly along a jagged gap that had been ripped out of the pipe's cement by a dragging anchor. The diver measured the scar with a steel ruler, calling out the measurements as he went for the benefit of the technicians on the surface: 'Two point nine centimetres. Three point zero centimetres. Three point two centimetres.' He took out a piece of chalk, drew a heavy line around the damaged section, and asked for a basket of tools to be lowered down to him. Then he and the supervisor launched into a detailed technical discussion of the nature of the damage, while Snoopy's camera went on panning slowly up and down the scar. Despite the distortion, the diver sounded so businesslike and so close that he might have been speaking from the room next door. Then he called sharply, 'Hey, Rocky.'

The camera panned quickly up to his helmet and Rocky, the supervisor, made a sudden, anxious sign to his assistant. 'You rang, sir?' said the supervisor in a plump, Jeeves accent.

'I don't arf fancy a snack,' said the voice from the helmet.

The supervisor exhaled deeply. 'We've tried that before,' he said. 'But the sandwiches seem to disintegrate and the salt's a bit of a problem. But hang about, they're working on it back at the beach.'

Snoopy began to pan regularly from the diver to the murk above him, up and back, up and back, like a dog anxiously looking for its master.

A thin, fidgety young diver, carrying a black leather instrument case, appeared at the door of the control room. The supervisor glanced at him briefly and said, 'What are you up to?'

'I'm trying my hand at the clarinet,' said the thin diver.

The supervisor turned back to the control panel.

'I used to play the flute,' the thin diver said to me. 'I took it into sat. with me and the first time I played it the controller came in on the phone and said, "All these years, I've never heard anyone whistle like that." '

The camera went on panning up into the dark water, then back to the diver. Finally, a hefty steel-mesh cage, painted white, dropped slowly down towards the pipe and settled on the seabed nearby, sending up a little cloud of sand. The diver moved over to it in slow motion, unlatched its door, and began to carry the gear across to the pipe: sonic measuring instruments, a stereo camera that looked like a sawn-off, double-barrelled anti-tank gun. As he moved methodically between the cage and the pipe, the supervisor said, 'Typical businessman going to work.' The diver guffawed into his mask. He took out his piece of chalk again and made more marks on the pipe. In his gloved hand the stick of chalk looked too small to use.

As I watched him, it occurred to me that I was taking the chalk for granted: what else would he use to make a mark on a damaged piece of pipe? Then I remembered that he was 470 feet below the surface of the water. What I was witnessing was a mystery. A mystery with a technical explanation, of course, but no less a mystery for that.

In a way, the Diving Control Room is a model for the whole offshore enterprise. The first time you see the banks of gauges with their flickering needles, the digital read-outs, the levers, the switches, the coloured lights, and, above all, the images on the video screens – one of them apparently full of seaweed and sunlight, with a pair of hands holding a book at its centre, like some obscure religious icon, the other with a helmeted spaceman moving in slow motion through the shadows where fish come and go indifferently – it seems like the greatest show on earth, a technological miracle, a justification of every boast about man's ingenuity and his ability to organise. But within an hour, you treat it as the people concerned treat it – as just another job, more complicated than most, more difficult than many. The great depth of cold, dark water is simply another obstacle to be overcome in the name of efficiency. Nobody mentioned the fear of the swim that David Hamm had said every diver felt, nobody mentioned the directionless darkness or the cold or the exhaustion or the overwhelming size of the structures down there under the water that made the divers feel as inconsequential as houseflies.

'Diving is demanding,' said Shane Beary. 'But it's nothing like as risky as, say, climbing. In the mountains you don't have the back-up we have, nor a certified training scheme. You could find yourself on a rope with any old fool, if you weren't careful. In diving nothing is left to chance. The highest accident rate is not in the water, it's in the home. Partly that's caused by the strain of the job, partly by being away for a month at a time, and partly by the sums we earn. Guys come out here from the services and suddenly find themselves deluged with money. They buy themselves expensive new toys – cameras, hi-fi, jewellery, a shiny new car. Finally, they get themselves a new wife. That's the real danger, not the diving. Diving is no more dangerous or romantic than working with high-voltage electricity. Provided you stick to the rules, follow the set procedures, and do what you're told, you won't get hurt. The trouble starts when you get back to the beach. When you've been living in this alien environment for four weeks on end, the world back home seems equally alien. One of the few releases from all that tension is a good, old-fashioned, forty-eight-hour piss-up. But try explaining that to the missus.'

12

In July above the sixty-first parallel, there is no night to speak of. At nine-thirty in the evening the pale sun still rode clear of the western horizon. Two hours later, the great circle of water and sky was seal-grey, not black. But no matter what the hour was, the video screens in the Diving Control Room were never blank, the gauges went on flickering, and the air was filled with the amplified rasp of the diver's breath, like a subtly irregular metronome measuring a different, submarine time: 'Huuuaaa . . . huuuaaa.'

Elsewhere on this man-made archipelago, spotlights glared on to the drilling floors where roughnecks heaved at the stands of pipe; far beneath the seabed, drilling bits turned steadily; roaring engines separated gas from oil, chippings from drilling mud; technicians were measuring, checking, taking notes, mending, relaying messages. And three hundred miles away in Aberdeen computers recorded every function of every working part. It was as though the whole gigantic enterprise had a strenuous life of its own that kept going no matter what the time, no matter what the weather.

With no night and the unceasing work – all that effort, all that attentiveness – it should have been hard to sleep. But the cabins on *Stadive* were spacious and comfortable, and the vessel swayed slightly, lullingly, as she shifted about above the divers or moved gradually forwards up the pipeline, yard by yard by yard. With the cabin door closed, the sound of the thrusters was reduced to a ghostly hum. But above that, distant yet distinct, was a faint, stuttering, intermittent creak, like the sound of someone typing fast in another cabin, then the long pause as his inspiration faded. I slept like a child every night and woke full of energy, although always a little uneasy at being the only person on board without a timetable and without a specific job to do.

I spent my last morning, as I had spent most of my time on *Stadive*, watching the divers on the video monitors. Once again they were piling bags of grout under an exposed span of pipe, routine work, simple and exhausting. After lunch, I went up on deck for a last look round. The day was cloudless and hot – shirtsleeve weather even this far north and in the middle of the sea. One of the diving bells had been brought up from the moon pool for repairs and a crew of technicians was swarming over it, inside and out, like wasps around a pot of jam. Two men were making adjustments to the MISS and another crew was at work inside the habitat. One of them was singing – a hollow, metallic sound, as though from the bottom of a well: 'I belong to Glasgee, dear ol' Glasgee toon . . .'

Flight 7, a Sikorsky S-61N, lifted off at four-fifteen sharp. Just before it arrived, a cheerful, beefy man called Mac, with a fiery complexion and nursery-blue eyes, who had greeted me affably every time I saw him in the corridors and the mess, appeared on the helideck wearing headphones under his hard hat. 'Wotcher doing here?' he roared. 'Running out on us?' He let out a great bellow of laughter and bustled off. For a moment, I wished I were staying. But only for a moment.

Going home, everything was the same as before, like a film run backwards: the chopper lifting, pausing, curtseying, then swooping gracefully forwards and up; the installations hazy in the milky green sea, with ranks of pipes like portcullises between their legs, their banners of flame streaming serenely into the sunlight. At Sumburgh, waiting for my connection, I drank watery coffee, then wandered outside to watch the spray rising and falling against a blue headland. At Aberdeen the airport was crammed with the same tough young men with tattooed forearms and oil company stickers on their duffel bags, and with the same grey-suited international businessmen – British, American, Dutch, Norwegian, French. But back at Heathrow the sky was overcast, thundery and hot. Indian cleaners moved listlessly through the crowd, pushing mops, while the sweating passengers complained about baggage that didn't come. Then the emptied faces in the tube, the bustle and dejection of the Big Smoke. It seemed a long way from the Shetlands with their voes and lakes and silence, their empty roads and cool air, even farther from the unremitting work and technological genius of the Brent field, and farthest of all from that single man at the end of his hose under five hundred feet of water, piling sandbags and measuring the damaged pipeline in the darkness.

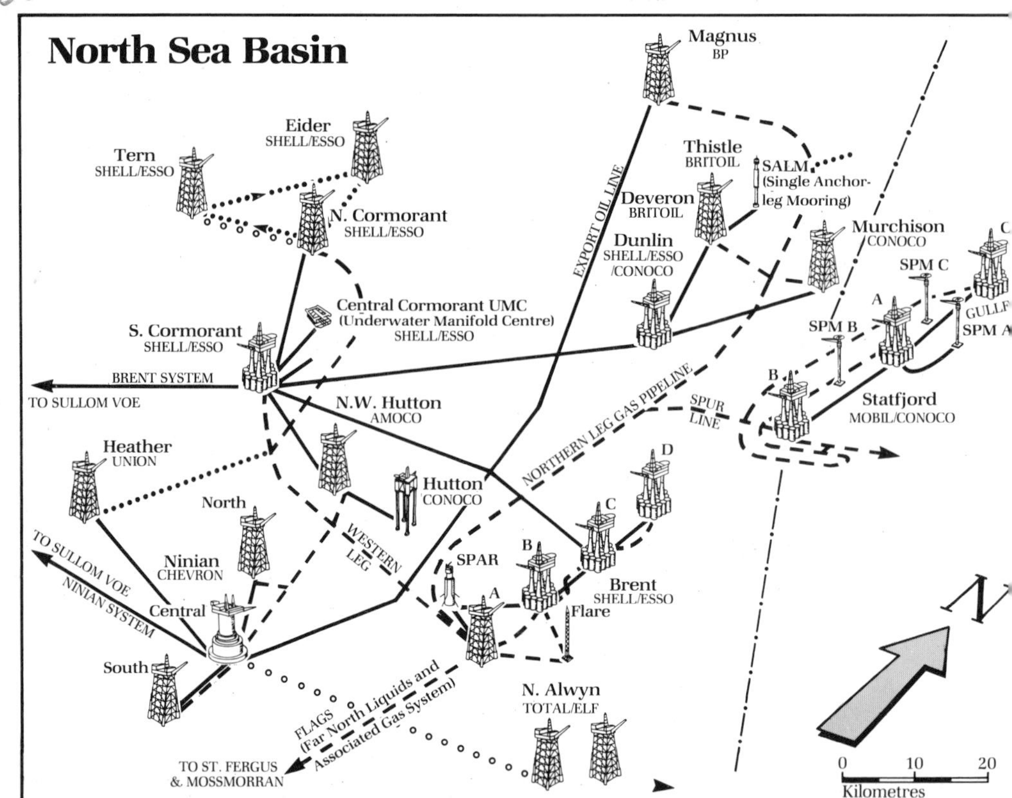
North Sea Basin
Magnus
BP
Eider
SHELL/ESSO
Tern
SHELL/ESSO
Thistle
BRITOIL
SALM
(Single Anchor-
leg Mooring)
Deveron
BRITOIL
N. Cormorant
SHELL/ESSO
EXPORT OIL LINE
Dunlin
SHELL/ESSO
/CONOCO
Murchison
CONOCO
SPM C
Central Cormorant UMC
(Underwater Manifold Centre)
SHELL/ESSO
S. Cormorant
SHELL/ESSO
A
SPM B
GULLF
SPM A
BRENT SYSTEM
TO SULLOM VOE
B
Statfjord
MOBIL/CONOCO
N.W. Hutton
AMOCO
NORTHERN LEG GAS PIPELINE
SPUR
LINE
Heather
UNION
Hutton
CONOCO
D
North
C
TO SULLOM VOE
NINIAN SYSTEM
WESTERN
LEG
B
Ninian
CHEVRON
SPAR
Brent
SHELL/ESSO
Central
A
Flare
South
N. Alwyn
TOTAL/ELF
FLAGS
(Far North Liquids and
Associated Gas System)
TO ST. FERGUS
& MOSSMORRAN
N
0
10
20
Kilometres

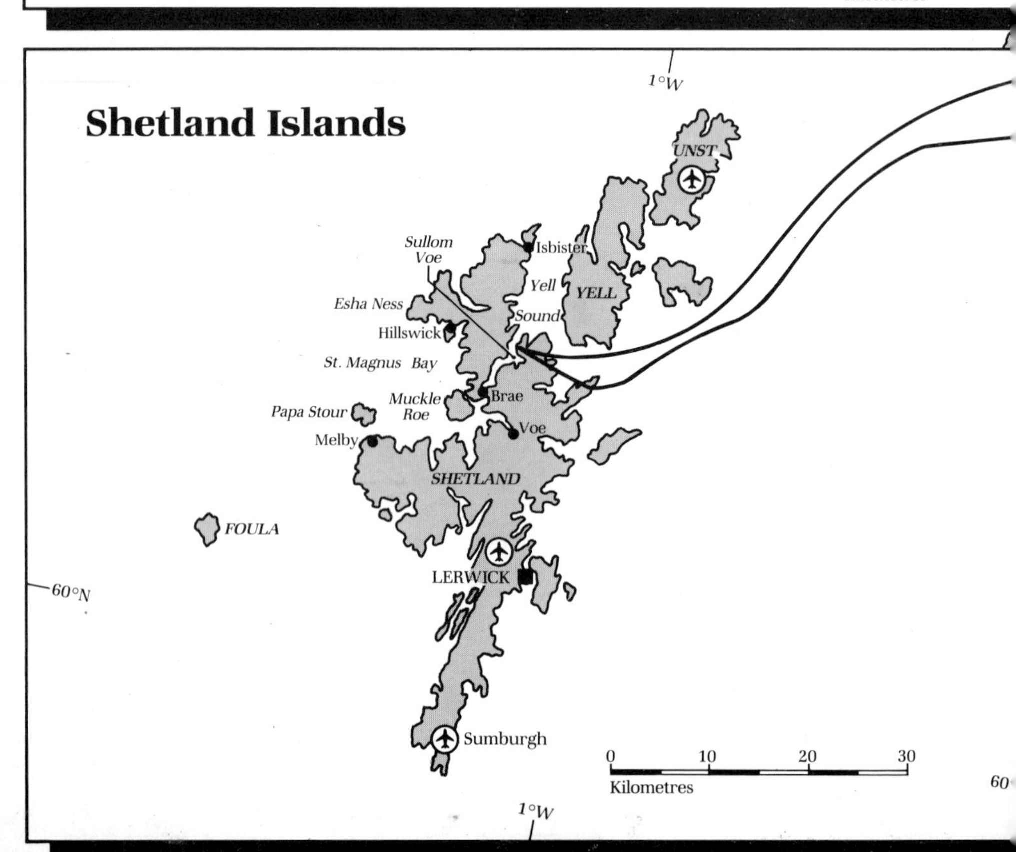
Shetland Islands
1°W
UNST
Sullom
Voe
Isbister
Yell
YELL
Esha Ness
Sound
Hillswick
St. Magnus Bay
Muckle
Roe
Brae
Papa Stour
Voe
Melby
SHETLAND
FOULA
60°N
LERWICK
Sumburgh
0
10
20
30
Kilometres
1°W